William Henry Craig

Doctor Johnson and the Fair Sex

A Study of Contrasts

William Henry Craig

Doctor Johnson and the Fair Sex
A Study of Contrasts

ISBN/EAN: 9783337398859

Printed in Europe, USA, Canada, Australia, Japan

Cover: Foto ©berggeist007 / pixelio.de

More available books at **www.hansebooks.com**

DOCTOR JOHNSON

AND

THE FAIR SEX

A STUDY OF CONTRASTS

BY

W. H. CRAIG, M.A.

Of Lincoln's Inn

(*WITH PORTRAITS*)

LONDON
SAMPSON LOW, MARSTON & COMPANY
LIMITED
St. Dunstan's House
FETTER LANE, FLEET STREET, E.C.
1895

ADVERTISEMENT.

To ALL WHOM IT MAY CONCERN.

My chief object in writing this book was to provide those who are curious on the subject, but lack time or inclination for research, with a convenient summary of what is known about the relations that existed between Dr. Samuel Johnson and divers notable women of his time. To this I have added short descriptions of certain ladies whose names, though occurring frequently in the story of Johnson's life, may have become more or less unfamiliar to the average reader of our own day. My facts are borrowed mainly from Boswell, Horace Walpole, Sir N. Wraxall, Madame d'Arblay, Mrs. Piozzi and Hannah More; but I have appropriated whatever suited my purpose elsewhere, without limitation as to authorship, whilst, through becoming lost in a wilderness of various editions, I have been less scrupulous in the matter of references than beseemed a compiler. Those, however, who wish to verify any of my statements can do so with the aid of Dr. Birkbeck Hill's noble edition of Boswell, wherein "chapter-and-verse" is given for well-nigh everything which has been recorded anent his favourite theme.

W. H. C.

LONDON, 1895.

CONTENTS.

	PAGE
I.—DR. JOHNSON AS A SQUIRE OF DAMES.	1
II.—DR. JOHNSON AS A SUITOR	32
III.—DR. JOHNSON AS A MAN OF FASHION.	70
IV.—DR. JOHNSON ON DRESS AND DEPORTMENT	187
V.—DR. JOHNSON ON MARRIAGE AND THE RELATIONS OF THE SEXES	203
VI.—DR. JOHNSON AS A KNIGHT-ERRANT	224

LIST OF ILLUSTRATIONS.

HEADPIECE—"THE POST-CHAISE" . . *page*   1
PORTRAIT OF DR. SAMUEL JOHNSON, AFTER
 REYNOLDS . . . . *frontispiece*
PORTRAIT OF MRS. ABINGTON . *to face page* 83
PORTRAIT OF MISS BURNEY . . ,, ,, 100
PORTRAIT OF MRS. CARTER . . ,, ,, 107
PORTRAIT OF MRS. CHAPONE . ,, ,, 114
PORTRAIT OF MRS. CLIVE . . ,, ,, 119
PORTRAIT OF MRS. DELANY . . ,, ,, 122
PORTRAIT OF MRS. GARRICK . ,, ,, 126
PORTRAIT OF MRS. KNOWLES . ,, ,, 130
PORTRAIT OF MRS. LENOX . . ,, ,, 134
PORTRAIT OF MRS. MACAULAY . ,, ,, 142
PORTRAIT OF MRS. MONTAGU . ,, ,, 157
PORTRAIT OF MISS HANNAH MORE ,, ,, 152
PORTRAIT OF MRS. PIOZZI . . ,, ,, 67
PORTRAIT OF MRS. THRALE . . ,, ,, 51
PORTRAIT OF MRS. SIDDONS . . ,, ,, 173
TAILPIECE—"THE LOST SHEEP" . . *page* 243

INDEX OF NAMES

	PAGE
ABINGTON, MRS.	3, 80–5
ADAMS, MISS	85, 86
ADEY, MISS	86, 87
ASTON, MISS	37–41
BABELS	136, 182
BARETTI	206, 228
BATHEASTON	75, 144–6
BLUE-STOCKINGS	73–8, 145, 157, 163
BOOTHBY, MISS	69, 87–90
BOSCAWEN, MRS.	91–4
BOSVILLE, MRS.	90, 91
BOSWELL, MRS.	94–8
BOSWELL, VERONICA	9
BROWN, MISS	192–3
BURNEY, MISS F.	65, 98–103, 126, 193
BURNEY, MRS.	193, 194
CARELESS, MRS.	33–5
CARMICHAEL, MISS.	22, 23, 234
CARTER, MRS.	29, 103–9, 198 *n*.
CHAMBERS, CATHERINE	241

	PAGE
CHAPONE, MRS.	92, 110–15
CHOLMONDELEY, MRS.	109, 110
CLIVE, MRS.	3, 19, 115–20
COBB, MRS.	86, 87
CORK, LADY. *See* MONCKTON.	
CORRICHATACHIN	8
DELANY, MRS.	120–2, 125, 158
DESMOULINS, MRS.	22, 23, 233, 234, 237
DEVONSHIRE, DUCHESS OF	6
EGLINGTON, COUNTESS OF	9
EMMET, MRS.	35, 36
FITZHERBERT, MRS.	29, 88, 122–5
FLOYD, MISS.	32
GALWAY, LADY	75, 147–8
GARRICK, MR.	36, 42, 82, 118, 128, 152
GARRICK, MRS.	125–8, 152
GASTRELL, MRS.	128–30
GLASSE, MRS.	18
HARVEY, MISS	131
HECTOR, MR.	33
HEELEY, MRS.	238
HOB-IN-THE-WELL	35
'IRENE'	118, 168

Index of Names.

	PAGE
JOHNSON, MRS.	40, 42–9, 219
JOHNSON, MRS. (SENIOR)	45, 239
KNOWLES, MRS.	27, 130–2, 201
LENOX, MRS.	108 *n.*, 132–4
LEVETT, MR.	23
LLOYD, MR. AND MRS.	210, 211
LUCAN, LADY	134–6
LYTTELTON, LORD	39, 89–90, 162
MACAULAY, MRS.	137–43
MCKINNON, MRS.	7
MAXWELL, REV. DR.	18
MEYNELL, MR.	29, 125 *n.*
MILLER, LADY	75, 143–6
MONCKTON, MISS	16, 74, 75, 146–51
MONTAGU, MRS.	73, 74, 99, 136, 156–63, 176, 177, 181, 182
MORE, MISS HANNAH	4, 74, 92, 127, 151–6
NORTHUMBERLAND, DUCHESS OF	76
ORD, MRS.	73, 163–5
PAOLI, GENERAL	209
PIOZZI, MR.	57, 64, 68
PIOZZI, MRS. *See* THRALE.	

Index of Names.

PAGE

PORTER, MRS. E. *See* JOHNSON.
PORTER, MISS L. 32, 43
PRITCHARD, MRS. 165-9

REYNOLDS, MISS 17, 29

SEWARD, MISS 18, 41, 170-2
SHERIDAN, MRS. 19
SIDDONS, MRS. 172-4
SKYE, ADVENTURES IN 6-15
STAFFORDSHIRE LADIES 18
STILLINGFLEET, MR. 73, 74

THRALE, MR. 39, 50, 51
THRALE, MRS. . 17, 27, 44, 49-69, 130, 133, 174-81,
205, 217
VESEY, MRS. 73, 79, 136, 176, 181-3

WALMSLEY, REV. G. 37
WILKES, MR. JOHN 10, 139
WILLIAMS, MRS. ANNA 22, 23, 52, 226-33, 236, 237
WILLIAMS, MISS HELEN MARIA . . . 183-6

"THE POST-CHAISE."

DR. JOHNSON AND THE FAIR SEX.

I.

DOCTOR JOHNSON AS A SQUIRE OF DAMES.

"IF I had no duties, and no reference to futurity, I would spend my life driving briskly in a post-chaise with a pretty woman." Such was the deliberate pronouncement of a philosopher verging on seventy; and, despite the ominous hint about futurity, it is surely

B

one of the finest compliments ever paid to the sex. What is more, it accurately represents, which few compliments do, the honest conviction of the speaker. We know from his own lips, and the testimony of his friends, that there were two things in which the Doctor's soul delighted—rapid motion, and the society of agreeable young women. Whirling along in a post-chaise was his notion of true enjoyment from a physical point of view; conversing with some sprightly beauty who could understand him and add something to the conversation, his acme of intellectual happiness. For this ordinarily uncouth and quarrelsome old man; this rampaging, brow-beating controversialist—who, at other times, betrayed a savage pleasure in flouting the amenities of social intercourse— could change himself into a vastly different monster when in the company of women,— could sheathe his claws, smoothe his bristles, and moderate his roar, when they patted and fondled him. What is stranger, he was always ready to forsake his predatory pursuits for the patting and fondling in question.

Dr. Johnson as a Squire of Dames. 3

All through life, from early boyhood to extreme old age, he exhibited this curious preference. Impelled by it, he would desert his accustomed haunts at club or tavern, where he domineered over lesser monsters with the unchallenged supremacy beloved by masterful natures, for those brilliant scenes where beauty and fashion were the ruling powers, and where his ungainly presence seemed an incongruity. He would hurry from his natural diet on books and papers to spend a quiet evening in copious tea-drinking with a few female friends. He would throw over Burke and Reynolds, Langton and Beauclerk, Goldsmith and Warton, to have a cosy chat behind the scenes with Mrs. Abington or sweet Kitty Clive. He would forego classics, criticism, philosophy, *belles-lettres*, to talk of caps and manteaux and other female gear with the gentle creatures to whom they were the main business of life. And why? Baretti, most incredulous of carpers, attributes this preference to a weak craving for adulation. "Almost every woman he (Johnson) knew," he remarks spitefully, "was a woman of uncommon merit with him, if they but coaxed him, which they all did."* But the good Doctor was not such a fool. That he did not object to flattery, in reason, is true enough; but he drew the line sooner than Baretti would have us believe, -and it is on record that Mrs. Hannah More incurred his severe displeasure by praising him too lavishly. This, by the way, leads us to the remarkable fact that women met the Doctor's advances at least half-way. They did not content themselves with simply receiving his homage, but they returned it. If he affected their society, they courted his with an assiduity that never flagged. Did he frequent their routs and assemblies? they in requital honoured his seclusion with angel visits *not* few or far between. Ladies of rank and fashion came in their sedan-chairs to the Temple, and in their fluttering robes tripped up its not too clean stairways, to visit the unkempt bookman at his grimy chambers. At festive gatherings they would quit their partners—those splendid

* 'Marginalia.'

Dr. Johnson as a Squire of Dames. 5

beaux of the Georgian times—to sit by Mr. Johnson in his shabby clothes and ill-trimmed wig. They adapted themselves to all his moods. If he happened to be dictatorial, as was not infrequently the case, they listened meekly. If, in a lighter vein, he vouchsafed them some of that genial flattery which he had always at command when he chose, they repaid him with compound interest. It was a fair game of give and take on both sides. "I am much pleased with a compliment from a pretty woman," owned the sage. "I love to sit by Dr. Johnson; he always entertains me," simpered the beauty. His physical infirmities, uncouth gestures and acerbities of temper only seemed to attract them; for, with that wonderful intuition which Heaven has granted women for their guidance, they soon divined that purest gold lurked beneath the rough quartz of his outer man. No doubt it was from this conviction that Johnson was petted and fondled and flattered by the women of his time to an extent that probably mortal man never was before or since. Wraxall describes how at the most fashionable assemblies he has seen, upon Dr. Johnson making his appearance, all the ladies present cluster round him in a circle four or five deep; and how he actually beheld the beautiful Duchess of Devonshire —Gainsborough's Duchess—then in the first bloom of youth, "hanging on the sentences that fell from Johnson's lips, and contending for the nearest place to his chair." Similarly, when he happened to go behind the scenes at Drury Lane and Covent Garden, as he pretty often did, there would be quite a flutter amongst the theatrical queens, who were also great social powers in those days; and happy was she who could get the Doctor to herself for a few minutes whilst waiting for the call-boy. Nor was this popularity confined to ladies of note, rank, or culture. A young woman of no particular pretensions once confided to Mr. Peter Garrick, brother of *the* Garrick, that, in her opinion, Dr. Johnson was "a very seducing man." Further, Boswell relates in his 'Tour to the Hebrides' how, when he and Johnson were disporting themselves in

Dr. Johnson as a Squire of Dames. 7

that *ultima Thule*, the simple, kindly Scotch dames whom they encountered actually lavished caresses upon their formidable visitor. Whilst they are in Skye, he reveals the fact that "one of our married ladies, a lively, pretty little woman, good-humouredly sat down upon Dr. Johnson's knee, and being encouraged by some of the company, put her hands round his neck and kissed him." The Doctor proved equal to the occasion : "Do it again," said he, "and let us see who will tire first." At another stage of their journey some artless Highland beauties became "eager to show their attentions to him, and vied with each other in crying out, with a strong Celtic pronunciation, 'Toctor Shonson, Toctor Shonson, your health!'" Nay, it is further recorded of a certain Mrs. McKinnon, the wife of a farmer or petty laird in the island of Skye, that while the punch went round of an evening she and the Doctor kept up "a close whispering conversation which, however, was loud enough to let us know that the subject of it was the particulars of Prince Charles's escape." This collogueing,

Boswell informs us, gave rise to some goodnatured banter. "We were merry with Corrichatachin on Dr. Johnson's whispering with his wife. She, perceiving this, humourously cried, 'I am in love with him! What is it to live and not to love?' Upon her saying something which I did not hear, or cannot recollect, he seized her hand eagerly and kissed it." The significance of all these innocent endearments lies in the primitive character of the worthy folk among whom they took place. Living where and as they did, these remote islanders could not have been much influenced by the glamour of literary achievements; and they must simply have taken Johnson as they found him, without regard to any estimate which the rest of the world had formed. And further, it is to be borne in mind that people living in such conditions are ever wont to form their impressions of a stranger's character from his outward appearance, from his looks and his way of carrying himself—which in Johnson's case were a poor letter of introduction. Yet here, and elsewhere in Scotland, we find that, despite his rugged and seamed lineaments, his convulsive movements, his untidiness, his loud harsh voice and "bow-wow manner," women and children took to him at once. Even Boswell's little daughter, Veronica, a timid shrinking child of eight, did not at their very first interview object to being held to the side of this queer old man—a proof of confidence which enraptured her father so much that he straightway resolved to add five hundred pounds to her "portion." And lest it should be fancied that this undisguised liking for their ungainly visitor was restricted to such Caledonians as were "stern and wild," mention shall be made here of the aged and stately Countess of Eglinton, to whom Johnson was presented during his stay in Scotland. That august lady, having learned in the course of their conversation that he had been born in the year after her own marriage took place, "graciously said to him that she might have been his mother: and when we were going away," adds the narrator, "she embraced him, saying, 'My dear son, farewell!'" Old and young, gentle and simple, all good women, all innocent children, were somehow drawn by a mysterious gravitation to the terrible Doctor.

Wherein lay the secret of this attraction? On the surface we can perceive that which might well account for it in some cases. His wonderful mental gifts, his literary reputation, his unrivalled power of conversation, and the fact that he was, so to speak, " the fashion," are sufficient to explain it, if we only take into consideration that class of women among whom Johnson habitually lived and moved—the politicians and scholars, the writers and readers; the women with pseudo-literary proclivities; and, above all, those who follow the trail of the lion for the time being, not because they love the chase, but simply because he is a lion. Due weight, too, must be allowed to the fact that women, as a rule, prefer the society of clever men, even when they do not happen to be themselves particularly clever. Though unable to shine, they love to be near the light. So that John Wilkes's boast of his power, ugly as he was, "if only given half-an-hour's start" to outrun the prettiest fellow in

England, where a lady's favour was the goal, is merely an illustration of a general law. For it requires no long experience of the sex to convince us that their regards are given, not so much to good looks as to the possession of manly qualities. Whatever we may think to the contrary, women are keenly alive to the fact that little of the good work which has been done in this world of ours is due to fine gentlemen with regular features; and before all things they put achievement. But, even when due allowance has been made for the influence which our good Doctor's reputation must have had with the classes of women above mentioned, the problem has yet to be explained, how he was such a favourite with those who had no love, or indeed knowledge, of letters—to whom literary reputation was an unmeaning sound—with the farmers' wives and the little children. To solve that we must fall back on the general question, why people are liked for themselves—that is, apart from their antecedents or surroundings? And all we know upon this subject is that there are people who, without being beautiful, or amiable, or moral, or respectable, or clever, or even cleanly, possess some mysterious quality which predisposes man and brute to foregather with them. Call it odic force, or animal magnetism, or sympathetic attraction, or by whatsoever high-sounding name we please, it is undeniable that some such force exists, and that its influence is most marked upon unsophisticated natures. Everybody can call up instances of its operation upon the men, women and children about him; he may even feel conscious of being more or less subject to it himself, and may certainly, if he choose, observe many wonderful exemplifications thereof amongst the cats, dogs and horses of his acquaintance. That Johnson possessed a more than ordinary share of this quality is the only way that occurs to us of explaining a very remarkable fact. It may not be scientific; indeed it may not be true, but it has at least some warrant in experience. And, granting the assumption, it seems to follow that in his case the quality must have existed in that somewhat rare form which inspires respect no less than liking. The respect moreover must have been instinctive; for what could those out-of-the-way Skye-folk have seen in his awkward shambling gait, his uncouth gestures, his seamed distorted features, his shabby garments, his want of the outer signs of wealth or position, to inspire respect? What could they have known of that piety, benevolence, charity, which had won him esteem elsewhere, any more than they did of the scholarship, authorship, or conversational pre-eminence, which had rendered him so famous? In all probability they could not have given any reason, even to themselves, why they were attracted to this strange old man, who had everything about him calculated to repel those who draw their conclusions from first impressions, and who lacked the only talisman by which such impressions are corrected among people in their circumstances— the power of displaying generosity. Yet the poor illiterate women, the shy impressionable children, took to him at once, followed him about, clung to him, caressed him. It is wonderful, it is inexplicable, except by some such wild hypothesis as we have suggested.

Through the dim glances of his bleared eyes, through the workings of his scarred features, they must by some God-given instinct, not by mere human reasoning, have discerned the beautiful soul within the man. Some divine voice must have whispered to them : " Here is one whom you may trust, who has great capacities of loving you, who has a kind and generous heart, a manly courage to protect, a pure intent to cherish and counsel you if need be, a charity to overlook your failings, a nobility to interpret your actions according to their true motives. Him you may safely trust and cling to; for he is strong, yet without guile." And we are not sure that to all this may not have been added an impression of his masterfulness—that quality which to weak natures is so attractive, and which even those women who are not of a weak nature prize above all things in a man, no matter what they may affect to the contrary. Every woman, every child, likes that which it can look up to, and which expects to be obeyed ; we doubt, indeed, whether, in the case of either, perfect love can exist without the apprehension of such a quality. But we need speculate no further on this subject; whatever the reason, the fact remains that, with women especially, Johnson's popularity was wide and strong. Adepts in mental physiology may bethink them of sounder and more practical ways of accounting for the phenomenon in question than what we have just propounded, and which is only offered for want of a better; but until they do so, it may serve to fill a gap left by all biographers of the worthy Doctor.

How did Johnson repay this reverential love and trust accorded him by a sex which ordinary men are wont to find imperious and exacting? Truth to say, in a manner which we cannot suppose to have been altogether satisfactory to beings of their sensitive organisation. Tenderness and affection he gave them freely enough; likewise that delicate flattery, conveyed not so much by language or demeanour as by undisguised preference for their society. But women ordinarily desire more than this; they very properly expect to be placed upon the same intellectual level with men, and this Johnson absolutely refused to do. He might own them to be clever, well-read, witty, and so forth; but he drew a marked and unflattering distinction between their mental capacity and that of the rougher sex. He never took them quite seriously, or affected to conceal from them his sense of their inferiority. Through the compliments of which he was so lavish there ever ran a vein of good-natured, bantering depreciation which, unless women have changed very much since his day, must have sorely tried the patience of those to whom they were addressed. Nay, too often he was rude—downright rude—when women ventured to propound opinions of their own; for, as one of his warmest admirers has sorrowfully recorded, "when Dr. Johnson met with opposition he respected neither age, nor rank, nor sex." Let us take a few examples of his "provoking" manner in this respect. When the sprightly Miss Monckton—she who afterwards became Countess of Cork, who was one of the leaders of fashion, and in whose society he much delighted—happened to remark that some of Sterne's writings were "very pathetic," Johnson bluntly contradicted her. "I am sure," pleaded the fair enthusiast, "that they affected *me.*" "Why," said he (smiling and rolling himself about), "that is because, dearest, you are a dunce." Mrs. Thrale, having once appeared at breakfast in a dark-coloured gown, was thus reproved— "You little creatures should never wear these sort of clothes," said her strange guest. "What! Have not all insects gay colours?" When Boswell was informing the company that he had lately been to a Quaker meeting where he had heard a lady preach, Johnson broke in with—"Sir, a woman's preaching is like a dog's walking on his hind-legs. It is not well done; but you are surprised to find it done at all." Miss Reynolds having asked his opinion of a metrical translation of Horace which a young lady had recently published, the reply came, "They are very well for a young Miss's verses—that is to say, compared with excellence, nothing; but very well for the person who wrote them." Nor would he even consent to admit that women could do things peculiarly within their province so well as men. "No, Madam," said he to Miss Seward, who had been praising Mrs. Glass's treatise on the culinary art, "women cannot make a good book of cookery"—at the same time assuring her that *he* could, if he chose. But perhaps the most nefarious instance of his contempt for the female understanding is that narrated by the Rev. Dr. Maxwell of the "two young women from Staffordshire" who had come to consult Johnson on the subject of Methodism, "to which they were inclined." How did the venerable sage respond to these earnest seekers after truth? "Come, you pretty fools," said he; "dine with Maxwell and me at the Mitre, and we will talk on that subject." Of course, the poor girls dared not refuse; so to the Mitre they all went, and it is sad to read that, "after dinner, he took one of them upon his knee and fondled her for half an hour together."

If the above excerpts show that Johnson's estimate of woman, as an intellectual force, was not very high, and that he had little compunction in avowing his opinion on the subject, it must not be inferred that he never took pains to conceal it; as a matter of fact, he prided himself on his gallantry. "I think myself a very polite man," he once remarked to Boswell; and it is beyond doubt that he could be very courtly in his behaviour to ladies when he chose. Upon such, unfortunately too rare, occasions there is reliable evidence that his kindness and easy pleasantry to young and old were delightful. Mention has already been made of the lady who declared him to be "a very seducing man"; and it was the famous Kitty Clive, by no means a tolerant critic of his sex, who said, "I love to sit by Dr. Johnson, he always entertains me." Nor could the smartest beau that ever figured at Almack's turn a compliment more neatly than the ponderous Doctor. When conversing with Mrs. Sheridan about her lachrymose novel, 'The Memoirs of Miss Sydney Biddulph,'—"I know not, Madam," quoth the artful sage, "that you have a right, upon moral principles, to make your readers suffer so much." The year before his death Mrs. Siddons, the actress, paid him a visit; and we are told that when she came into the room there happened to be no chair ready for her, which he observing, said, with a smile, "Madam, you who so often occasion a want of seats to other people will more easily excuse the want of one yourself." Boswell relates how the old man, at an evening party, once asked "an amiable, elegant and accomplished young lady" to sit down by him, which she did; and upon her inquiring how he was, he answered, "I am very ill indeed, Madam. I am very ill, even when you are near me: what should I be, were you at a distance?" Nor must that story be omitted of the young lady who, having just handed him some coffee, remarked that the coffee-pot was "the only thing she could call her own." "Don't say so, my dear," entreated the incorrigible Doctor, "I hope you don't reckon my heart as nothing." Pretty well this for an old gentleman who was then tottering on the very brink of the grave![*] And if, owing to his complex physical infirmities, he was precluded from offering ladies those attentions

[*] It was uttered in June, 1784; and he was buried in December following.

Dr. Johnson as a Squire of Dames. 21

which it is the privilege of younger and more agile men to bestow, there is not wanting evidence of his native gallantry in this respect. When the celebrated Madame de Boufflers, during her stay in England, honoured him with a visit at his chambers in the Temple, Boswell describes how, the interview being ended, "She and I left him, and were got into Inner Temple Lane, when all at once I heard a noise like thunder. This was occasioned by Johnson who, it seems, upon a little recollection, had taken it' into his head that he ought to have done the honours of his literary residence to a foreign lady of quality; and eager to show himself a man of gallantry, was hurrying down the staircase in violent agitation. He overtook us before we reached the Temple Gate; and brushing in between me and Madame de Boufflers, seized her hand and conducted her to her coach." Perhaps an equally convincing, and far more touching, proof of his chivalry is that when on the very day of his death, a young lady begged to be presented to him, rather than disappoint the girl the old man consented to receive her;

and turning himself round in his bed as she entered the room, just managed to say—"God bless you, my dear!" They were the last words he spoke.

But it was to women overtaken by poverty, weakness, or misfortune, that the best side of Johnson's character revealed itself. To these he was never blatant or imperious; and so far as might be, he shifted their burdens to his own broad shoulders. No sooner had he got a house of his own than he packed it with what Boswell called his *seraglio*; but the Doctor's seraglio was not composed of "almond-eyed and moon-hipped" houris. Aged gentlewomen who had known better days—Mrs. Williams, the blind and querulous friend of his deceased mother; Mrs. Desmoulins, the widowed daughter of his godfather (Dr. Swinfen), together with her daughter, and the slatternly "Polly" Carmichael—were the sultanas who lodged beneath his roof, ate his bread, and shared his meagre purse.* To

---

* Boswell mentions that Mrs. Desmoulins herself told him Johnson allowed her half-a-guinea a week— more than a twelfth part of his pension.

them was added old Mr. Levett, a broken-down apothecary; and between them all they contrived to make Johnson's house anything but a temple of peace. In a letter to Mrs. Thrale he thus describes how they got on—"Williams hates everybody; Levett hates Desmoulins and does not love Williams; Desmoulins hates them both; and 'Poll' loves none of them." Nor did their benefactor come off scatheless amid the general hostility; albeit, he took his buffets without a murmur. Mrs. Williams might be peevish and exacting when she chose—and she chose it very often; but he, so quick to turn and rend others upon the slightest whisper of revolt, never answered her in kind, and strove to propitiate her by sending in goodies from the cookshop when he ventured to dine out. Weakness was all-powerful with this strangely constituted man. Even vice itself overcame him when it took that shape. For have we not that beautiful story of how, "Coming home late one night, he found a poor woman lying in the street, so much exhausted that she could not walk; he took her on his back, and carried her to his house, where he discovered that she was one of those wretched females who had fallen into the lowest state of vice, poverty and disease. Instead of hastily upbraiding her, he had her taken care of with all tenderness for a long time, at a considerable expense, till she was restored to health, and endeavoured to put her into a virtuous way of living." Surely a more Christ-like action was never done by man since the Good Shepherd walked this earth! As we think of the burly giant carrying that poor outcast through the dark streets the memory of his imperfections dies away, and the light of a better world than ours shining on his rugged features makes them beautiful.

It must be confessed that we are precluded from dwelling at any great length upon the amiable side of Johnson's character by sheer paucity of materials for so pleasant a task. The weight of evidence preponderates in the other scale of the balance which an impartial biographer is bound to use, and that allotted to his recorded suavities and affabilities already kicks the beam. Though it is beyond dispute that the man's nature overflowed with benevolence and the other Christian virtues, he cannot fairly be credited with having made a lavish display of these winning qualities in his daily conversation, or indeed in his writings. To borrow a metaphysical distinction, his soul was noble and loveable beyond expression, while his mind, despite its power and acuteness, displayed a captious, critical, hole-picking tendency which generally inspired fear, if it commanded admiration. To this tendency must, no doubt, be attributed the rampant heresy which blemishes his doctrine of the sexual relations. Eager to detect a flaw in the perfection of the female character, he did not hesitate to publish and avow as his firm conviction that woman was a being in many respects inferior to man, and therefore rightly held in that state of comparative subjection from which the sex are now so rapidly emerging. As might be expected, his logic is sometimes at fault when he attempts to establish this dogma; as for instance when, in writing to Dr. Taylor, he propounds the contradictory assertion—"Nature has given women so much power that the law has very wisely given them little." Or when, as in the *Rambler*,* whilst compassionating the disabilities under which women laboured owing to circumstances beyond their control, and pointing out that Nature has so ordered their lot that, whether married or single, they are bound to suffer, together with the evils of the forced marriages then in fashion, he goes on to say—"I have indeed seldom observed that when the tenderness or virtue of their parents has preserved them from forced marriage, and left them at large to choose their own path in the labyrinth of life, they have made any great advantage of their liberty. They commonly take the opportunity of independence to trifle away youth, and lose their bloom in a hurry of diversions recurring in a succession too quick to leave room for any settled reflection. They see the world without gaining experience, and at last regulate their choice by motives trifling as those of a girl, or mercenary as those of a

* *Rambler*, No. 39 (31 July, 1753).

Dr. Johnson as a Squire of Dames. 27

miser." If these words had been written before Mrs. Thrale formed the acquaintance of her second husband, we might suspect that they embodied a personal mortification and simply referred to an individual case ; but it is to be feared that this loophole of escape is closed to the misguided Doctor, as the Piozzi match was not yet on the carpet, and the general tenor of his writings and sayings points only too distinctly the other way. "Women," said he to Mrs. Knowles, " have all the liberty that they should wish to have : we have all the labour and danger, and the women all the advantage." And in another place he attributes the fact that women are more genteel than men, to that of their being "more restrained." He even denies them the power of moral judgment. "Ladies," in his opinion, "set no value on the moral character of men who pay their addresses to them ; the greatest profligate will be as well received as the man of greatest virtue, and this by a good woman—by a woman who says her prayers three times a day." Nay, he disputes their capability of choosing their own frocks properly.* "They are the slaves," he alleges, "of order and fashion," and care not for what is graceful and becoming. "Women in general have no idea of grace, and fashion is all they think of." Worse still, the unhappy Doctor, in what we presume to have been a paroxysm of temporary derangement, declares that "Women have a perpetual envy of our vices; they are less vicious than we—not from choice, but because we restrict them"! After this, one is not unprepared to find him writing †—"Gluttony is, I think, less common among women than among men . . . . but if ever you find a gluttonous woman, expect from her very little virtue."

We have here simply the outcome of that spirit of contradiction which distinguished Johnson all through life. Whenever he found an opinion he attacked it; and the mere fact that the general judgment of mankind tends

* Boswell assures us that the women with whom the Doctor was acquainted agreed that "no man was more nicely and minutely critical in the elegance of female dress."

† To one of Mrs. Thrale's daughters.

Dr. Johnson as a Squire of Dames. 29

in the opposite direction was sufficient to make him adopt, or pretend to adopt, the illiberal theory of sexual relations which we have briefly described. That it does not express his real estimate of women is tolerably apparent from our finding him every now and then marking the most unqualified exceptions from it. Thus, of one lady, Mrs. Fitzherbert, the gifted daughter of his old friend Meynell, he is reported to have said, "she had the best understanding that he ever met with in any human being." Of another, Miss Reynolds, sister of the painter, that her mind was "very near to purity itself."

In discussing the classical attainments of a contemporary, he remarked that the scholar in question understood Greek better than anyone he had ever known, "except Elizabeth Carter." Further, he made the, for him, very remarkable admission that after he had long puzzled himself over that not particularly recondite problem, "Why the interest of money is lower when money is plentiful?" it was solved for him readily by a youthful member of the sex whose judgment he had so often decried. Examples of this inconsistency need not be multiplied here; they will be found to crop up readily enough as we follow him into society, and the indignant fair may take for granted that the heresy which he so ungallantly professed was mainly "from the teeth outwards."

In other respects he was not unjust to women: he admitted their superior refinement and gentility, and he deplored the injustice of their seclusion from so many avenues to advancement. He declared that when they were given a fair chance of improving themselves they took full advantage of it: pointing out, as a proof of this, that the ladies of modern times were better in every respect than their predecessors, "because their understandings were better cultivated," though with the reservation, "when it comes to *dry understanding*, man has the better." That "there are ten genteel women for one genteel man" was among his axioms. "If we require more perfection from women than from ourselves," he protested, "it is doing them honour." He considered that the higher women rose in the social scale the better they became—which he by no means admitted to be as often the case with men; though, as a general rule, "high people, Sir, are the best." "Take a hundred ladies of quality," he once said; "you'll find them better wives, better mothers, more willing to sacrifice their own pleasure to their children, than a hundred other women." In fine, amid those wanton expressions which, from time to time, his cantankerous failing led him to utter, may easily be discovered the presence of a genuine admiration, emphasised by the eagerness with which, all his life long, he courted the privilege of women's society.

II.

Dr. Johnson as a Suitor.

Like many another spirited young fellow, Johnson loved early and loved often. Whilst yet a schoolboy he formed sundry evanescent attachments—one for Miss Olivia Floyd, "a young Quaker, to whom he wrote a copy of verses." Another is said to have been for Miss Lucy Porter, the daughter of his future wife, whom he honoured in a similar way "on her presenting him with a nosegay of myrtle"; though there is some doubt as to the genuine nature of this passion. Anyhow, he thus beseeches the lady :—

> " O then, the meaning of thy gift impart,
> And ease the throbbings of an anxious heart !
> Soon must this bough, as you shall fix his doom,
> Adorn Philander's head, or grace his tomb."

The throbbings were not eased by Miss Lucy, yet Philander survived to marry the parent of the charmer who was to have fixed his doom. In fact, it is admitted that "his juvenile attachments to the fair sex were very transient," though we are glad to find that none of them were unworthy. Boswell was assured on the authority of Mr. Hector, a gentleman who had lived with Johnson in his younger days in the utmost intimacy and social freedom, that "even at that ardent season his conduct was strictly virtuous in that respect."

It was perhaps natural enough that Johnson's first serious affair of the heart should have for its object the sister of this very Mr. Hector. Whilst he and Boswell were at Birmingham the Doctor observed one morning to the latter—"You will see, Sir, at Mr. Hector's, his sister, Mrs. Careless, a clergyman's widow. She was the first woman with whom I was in love. It dropt out of my head imperceptibly; but she and I shall always have a kindness for each other." Accordingly, at Mr. Hector's on the same day, Johnson is found sitting placidly at tea with his first love, "who, though now advanced in years, was a genteel woman, very agreeable and well-bred." Alluding to this same interview in a letter to Mrs. Thrale, the Doctor writes, " I have passed one day at Birmingham with my old friend Hector—there's a name for you—and his sister, an old love. My mistress is grown much older than my friend—

> ' *O quid habes illius, illius,*
> *Quæ spirabat amores,*
> *Quæ me supererat mihi.*' "

The old man was evidently touched by this meeting. When alone with Boswell that night, his affection for the lady seemed to have revived ; for he said, " If I had married her it might have been as happy for me." Why the match did not take place, there is nothing to show. Perhaps the love was only on one side ; but it is more likely that Johnson's circumstances at the time stood in the way, for we find a reciprocal complaisance between the aged couple which lends colour to the idea that their attachment was mutual. In his correspondence with her brother he invariably alludes to her as " dear Mrs. Careless," sometimes asking for her prayers ;

and in describing a subsequent meeting with her at the house of some friends, he says, "Mrs. Careless took me under her protection *and told me when I had tea enough.*" A hint of tenderness is conveyed in the authority which the lady chose to exercise over that insatiable tea-drinker. As it was not *her* tea which the Doctor was swilling on the occasion, her motive in restraining him can scarcely have been other than affectionate ; but this is pure matter of conjecture, and she may have thought him expensive—who knows? It is highly probable that another passion, of which we have only a brief record, should be referred to the period of his early youth, though the Doctor himself has left a rather compromising statement to the contrary. During their visit to Lichfield, which took place in 1776, he remarked in confidence to Boswell— " Forty years ago, Sir, I was in love with an actress here, Mrs. Emmet, who acted Flora in 'Hob in the Well.' "* Now, as he was

* A play of Cibber's which appears to have held the stage for some time, and occasionally figures in the bills of Covent Garden and Drury Lane theatres towards the end of last century.

born in 1709, he would, if this statement were correct, have been twenty-seven when the attachment in question was formed; but at that age he was a married man, and all that we know of his character is entirely opposed to the idea of his having ever been, even in thought, an unfaithful husband. There is little doubt, therefore, that the good Doctor was not sufficiently precise in fixing the interval which had elapsed since his acquaintance with Mrs. Emmet, or that perhaps another ten years might have been added to the "forty" which he specifies. Of the lady herself no details are forthcoming. Her name does not appear among those who have distinguished themselves in her profession—most probably she changed it, or quitted the stage, soon after making a conquest of his affections. Even the curious Boswell admits—"What merit this lady had as an actress, or what was her figure, or her manner, I have not been informed; but if we may believe Mr. Garrick, his old master's taste in theatrical merit was by no means refined: he was not an *elegans formarum spectator*. Garrick used to tell that Johnson said of an actor, who played Sir Harry Wildair at Lichfield, 'there is a courtly vivacity about the fellow'; when in fact, according to Garrick's account, 'he was the most vulgar ruffian that ever went upon boards.'" Be that as it may, there is nothing to show how long the flirtation lasted, or how far it went—probably no further than mute homage from the body of the house to Johnson's goddess for the time being.

So far, none of Johnson's love affairs had been very serious. We gather this from his free way of discussing them; for, though an outspoken man, he did not wear his heart upon his sleeve. But we now light upon an episode of his early days which, however faintly indicated, seems to contain the elements of a romance. Among the friends of his youth at Lichfield was the Rev. Gilbert Walmsley, Registrar of the Prerogative Court, who, though much his superior in age and station, delighted in the lad's society. At this gentleman's house, where he passed much of his time, he met Miss Molly Aston, daughter of Sir Thomas Aston, Bart., and sister-in-law of Mr. Walmsley. The baronet's daughters were all remarkable for their good breeding, whilst Johnson, as a young man, was "distinguished for his complaisance." The lady was both beautiful and attractive—we learn that she was not popular with her sex—and seems to have been clever as well.

Johnson once remarked to Boswell, " Kames is puzzled with a question that puzzled me when I was a very young man :—Why is it that the interest of money is lower when money is plentiful; for five pounds has the same proportion of value to a hundred pounds when money is plentiful as when it is scarce ? A lady explained it to me. ' It is,' said she, 'because when money is plentiful there are so many more who have money to lend that they bid down one another. Many have then a hundred pounds ; and one says, " Take mine rather than another's, and you shall have it at *four per cent.*"'" BOSWELL : " This must have been an extraordinary lady who instructed you, Sir. May I ask who she was ? " JOHNSON : " Molly Aston, Sir, the sister of those ladies with whom you dined at Lichfield."

That Miss Aston entertained any warmer feeling than friendship for the awkward lad is very unlikely. Poor Johnson was not an eligible *parti* in any one respect; he was also her junior, and his literary reputation was yet to come. But the young fellow's admiration of her seems to have been intense, and it lasted all through life. Honour forbade his breathing a word of passion to the relative of his wealthy benefactor; nor did he ever confide to anyone else that he loved her. But to those who could read between the lines the secret of his attachment was evident enough; and it was even said that certain unfavourable views expressed of Lord Lyttelton in his 'Lives of the Poets' were owing to jealousy excited by his lordship's attentions to Miss Aston. Mr. Thrale having once asked him what had been the happiest period of his past life, his answer was: "The year in which I spent one whole evening with Molly Aston. That, indeed, was not happiness, it was rapture; but the thoughts of it sweetened the whole year." "Molly," he remarked again, "was a beauty and a scholar and a wit and a *Whig* ... she was the loveliest creature I ever saw"; to which he added the curious information that "the ladies never loved Molly Aston." Also, Mrs. Piozzi narrates how he told her that Molly's letters would be the last papers he should destroy; and as no trace of them was found after his decease, it is probable that the old man burned those precious relics when he felt his time had come. Eventually, Molly became the wife of Captain Brodie, a naval officer; and we lose sight of her. But we find that Mrs. Johnson was jealous of her, though without any cause. And once, by a most unfortunate accident in her presence, a gipsy having told him that his heart was divided between a Betty and a Molly, adding: "Betty loves you best; but you take most delight in Molly's company," there was a painful scene. We also know that the "lady of great beauty and elegance" mentioned in his criticism upon Pope's Epitaphs was none other than Molly Aston.*

As we have seen, her image clung to his memory all through life, and such tokens of her friendship as he possessed were treasured by him to the end. He must have liked her

* Mrs. Piozzi's 'Anecdotes of Johnson.'

Dr. Johnson as a Suitor. 41

very much indeed when not even the "Whiggery" which she professed marred their intimacy; for we learn that, in answer to her high-flown speeches for liberty, he addressed to her the following epigram :

"*Liber ut esse velim suasisti pulchra Maria,
Ut maneam liber pulchra Maria vale.*"

But the "*vale*" was merely rhetorical; he had no thought of quitting such pleasant company, and for once in his life acknowledged that a Whig needs not be "vile." It is therefore natural to suppose that, when the separation came, it must have caused him many a bitter pang; though the probability is that he soon formed another attachment, for we have it on good evidence that "he laughed at the notion that a man never can be really in love but once, and considered it as a mere romantick fancy." Indeed, somewhere about this time there was rumour of a flirtation with the respectable Miss Seward; but it rests upon no solid foundation, and never came to anything.

However, when he had arrived at the fulness of manhood—being twenty-four, or thereabouts

—he began to take the disposal of his affections seriously in hand; and this time formed a really substantial, workmanlike attachment. There happened to be just then at Lichfield a lately-bereaved widow of matured charms— a Mrs. Porter—whose husband had been an acquaintance of his. The lamented Porter having been unfortunate in business—he was a mercer—died insolvent; but his relict, who was protected by her settlement, possessed a good deal of ready money, which was the very commodity that Johnson stood most in need of. Mrs. Porter was twice his age; and Garrick says that she was "rather a fright— very fat, with a bosom of more than ordinary protuberance; with swelled cheeks of a florid red, produced by thick painting and increased by the liberal use of cordials; flaring and fantastic in her dress, and affected both in her speech and her general behaviour." This portrait may, or may not, have been a caricature—most likely it was; but at all events her appearance captivated Johnson, who never alluded to it but in terms of the warmest admiration. What his own was at the time

Dr. Johnson as a Suitor.

may be gathered from the following sketch given by Boswell :—

"Miss Porter told me that when he was first introduced to her mother his appearance was very forbidding. He was then lean and lank, so that his immense structure of bones was hideously striking to the eye, and the scars of scrophula were deeply visible. He also wore his hair, which was straight and stiff and separated behind; and he often had, seemingly, convulsive starts and odd gesticulations which tended to excite at once surprise and ridicule."

But, as usual, and in spite of drawbacks, Johnson triumphed over the sex ; for we read how "Mrs. Porter was so much engaged by his conversation that she overlooked all the external disadvantages, and said to her daughter, 'He is the most sensible man that I ever saw in my life.'" Fortunately for both, she was inclined to literature and had a good judgment in such matters ; * otherwise, it is unlikely that she should have made a con-

---

* This is shown by her remarks on some of Johnson's writings, by which she proved herself quite capable of discriminating between what was merely good and what was better.

quest of the man who, when he declared that he could pass his life in a post-chaise with a pretty woman, stipulated that "she should be one who could understand me, and would add something to the conversation." There is, further, no ground for supposing that Johnson's appearance impressed her unfavourably —perhaps, like her suitor, the lady may have suffered from defective vision; whilst, as regards her own, Mrs. Thrale had seen a portrait of her which, she said, was "pretty." At all events, each had qualities to attract the other; for, as Johnson used solemnly to aver, "It was a love match, Sir, on *both* sides." Unfortunately, we are left without any record of the procedure which he observed in the conduct of his suit; but we may be sure that his methods were carefully thought out beforehand, and dexterously carried to their logical conclusion. Of course, no one save himself could frame the diction wherein his proposals were conveyed; but from our knowledge of his demeanour to winsome females in analogous circumstances, we can form some idea of the actions that accompanied his words.

It is almost certain therefore that when the proper moment arrived he kissed the lady's hand, or possibly her cheek ; and that, if she returned the salute, he requested her to "do it again." That he knelt to her when urging his suit is unlikely; but there is nothing at all improbable in the conjecture that he may have taken her upon his knee and "fondled" her for—say, "half-an-hour together." And of this we may be quite sure, that he would have been prepared to justify by argument any familiarity to which she might have objected.

Like a dutiful son, no sooner had Johnson ascertained the fair widow's readiness to accept him than he applied for his mother's consent to the marriage which, as we are told, "he could not but be conscious was a very imprudent scheme." Old Mrs. Johnson, however, "knew too well the ardour of her son's temper, and was too tender a parent to oppose his inclinations." Preliminaries being thus settled and the decks, as it were, cleared for action on both sides, nothing remained but to carry out the engagement. For some unexplained reason the marriage was fixed to take place—not at Birmingham, where the lady resided, but at Derby; and thither one fine July morning the pair set out on horseback. Discarding local custom, the blushing bride did not occupy a pillion behind her future lord, but had a palfrey all to herself. How they sped may best be told in Johnson's own words :—

"Sir, she had read the old romances, and had got into her head the fantastical notion that a woman of spirit should use her lover like a dog. So, Sir, at first she told me that I rode too fast, and she could not keep up with me; and when I rode a little slower, she passed me, and complained that I lagged behind. I was not to be made the slave of caprice; and I resolved to begin as I meant to end. I therefore pushed on briskly, till I was fairly out of her sight. The road lay between two hedges, so I was sure she could not miss it; and I contrived that she should soon come up with me. When she did, I observed her to be in tears."

Thus did Johnson lay a solid and workmanlike foundation whereon to rear the complex structure of married life; but, somehow or other, the result did not prove to be anything out of the common. For we find that when Mrs. Thrale once asked him if he ever disputed with his wife, the answer was, "Perpetually." It is interesting to learn that the chief subject of marital variance was a question of æsthetics. Mrs. Johnson happened to be passionately fond of cleanliness; her husband's proclivities were all the other way. Every married man knows how the controversy was likely to end—but we are digressing.

Did Johnson kiss away the lady's tears? The operation would have been a difficult one on horseback; yet, he was quite capable of attempting it. At all events they must have made up their differences by the way, for the register of St. Werburgh's Church, Derby, contains the following entry:—

"1735. July 9. Married Saml. Johnson of ye parish of St. Mary's in Lichfield, and Elizabeth Porter of ye parish of St. Philip in Birmingham."

In addition to valuable experience, and two grown-up children by her former marriage, Mrs. Johnson brought her husband a fortune of £800, which he forthwith utilised in setting up a boarding-school for young gentlemen at Edial, near Lichfield. The venture did not succeed; and after twenty months of wedded rapture Johnson had to forsake his wife temporarily, with a view to trying his fortune in London, whence, in the summer of 1737, he returned to Lichfield with little but 'Irene' in his pocket. Having once tasted the charms of the metropolis, those of the country were no longer an attraction; and, after a three months' stay at Lichfield, he removed with his wife to London, where for some time the pair lodged "in Woodstock Street, near Hanover Square, and afterwards in Castle Street, near Cavendish Square." Mrs. Johnson proved an appreciative, but not exactly what people call a "devoted" wife. While her spouse was labouring at the oar in some publisher's ill-found galley, she took her ease at home, like a sensible woman, with full content and pleasant cheer. Nay more, Boswell asserts that "she indulged herself in country air and nice living at Hampstead while her husband was drudging in the smoke of London; and by no means treated him with the complacency which is the most engaging quality in a wife." Yet his love for her continued to be of the most ardent kind; and when, after some sixteen years spent together, she died in 1752, his grief was most intense and lasting—a grief too sacred to be dwelt on here.

So great is the conflict of opinion as to the precise nature of Johnson's feelings towards the widow Thrale that a judicious biographer must hesitate about venturing to include that lady in the list of his "loves." It is therefore only as a tentative and provisional arrangement that her name, which will afterwards appear in the category of his "friendships," where it has an undoubted right to be, is made to figure here. Nor will the present writer propound any views of his own upon the question at issue. His function it will be to lay certain premises before his readers, who may thence draw such conclusions as please them best—a procedure by which he hopes to avoid the slightest imputation of scandal-mongering or impropriety.

In weighing the evidence upon this very delicate question it should never be forgotten that the relation towards each other into which the two principals were brought by sheer force of circumstances could not, in the very nature of things, have been one of mere ordinary friendship. Some years after the death of his wife Johnson made the acquaintance of the Thrales; that acquaintance ripened into intimacy, and the upshot was that "at last he became one of the family, and an apartment was appropriated to him both in their house at Southwark and in their villa at Streatham." With them he continued to live upon the most intimate footing for some sixteen or seventeen years, until Mr. Thrale's death. He was treated by them both with absolute unreserve, knew all their private affairs, was consulted upon every difficulty that arose, no matter of how confidential a nature; and was in fact, next to Mr. Thrale himself, the ruling authority in their household. Her he saw at all times and seasons; they talked together with perfect freedom upon all manner of subjects; she gave him

MRS. THRALE.

her confidence and invited his. It was as if he were her father—only more so; as if she were his daughter, but with a pleasant difference. Mr. Thrale, who was a heavy sort of man, used to fall asleep after dinner, and leave the trouble of conversation to Johnson and his wife—which might seem injudicious if the worthy Doctor had not been a most honourable gentleman and a sexagenarian to boot. When Johnson first made her acquaintance Hester Thrale must have been a very charming young woman. Having married only two years previously, whilst still almost a girl, she retained that bloom of which youth, perfect health, and a cheerful temper are the best preservatives. Her features would have been beautiful but for a slight blemish in the lower part of the face, caused by an accident. Miss Burney, on meeting her a full dozen years afterwards, declared that even then she was a pretty woman—" her nose is very handsome, her complexion very fair, and her eyes are blue and lustrous." If her figure lacked the willowy grace of girlhood, it possessed that substantial comeliness which wears so much better. Boswell, who did not like her, says she was "short, plump, and brisk," by which downright adjectives we are to understand that she was what the French call *petite*, her figure rounded and lissom, and her movements full of sprightly grace. In fact she was good-looking, good-natured, and good-tempered enough to please anybody who was not determined to find fault. To Johnson, fresh from the unlovely and fretful company of poor Mrs. Williams, she must have seemed positively enchanting. And then, she was so kind to the old fellow! She hovered about him perpetually with her bright sunny ways; she sat up with him patiently of nights, when, loth to seek that couch which, owing to the sleeplessness and the other infirmities from which he suffered, inspired him with a kind of terror, the sage could not tear himself from the cheerful hearth until long after Mr. Thrale and the household in general had retired. She studied those creature-comforts which, philosopher as he was, he loved so well. She amused him with her sparkling wit and vivacity; listened to his wondrous talk with

Dr. Johnson as a Suitor. 53

unfeigned interest, and possessed an acquaintance with general literature which enabled her to take a responsive part in the conversation. So that in all three respects she fulfilled the conditions required of the woman with whom he declared that he could pass his life in a post-chaise—she was "pretty," she could "understand" him, and she could "add something to the conversation." What wonder, then, if the tender-hearted sage was charmed? Remember, he was then only fifty-six, which is the mere boyhood of old age. Did not the rigid and awesome Master John Knox—he who unflinchingly denounced "The Monstrous Regiment of Women"—in his sixtieth year succumb to wedlock with a chit of sixteen? Did not the sapient *von* Gentz, the subtlest intellect in Europe, at the age of eighty, woo, win, and run off with Fanny Elsler, who was then a budding *coryphée* of seventeen years? Man is, in fact, never too old to like a pretty woman ! Even in our ashes live their wonted fires ; and he must be in sooth a dull and stricken dotard in whom the presence of a young, blooming and agreeable member of the other sex fails to kindle some glow of admiration—we do not say of love, for, between ourselves, elderly gentlemen are rather out of place in that galley; nor do we recommend them to follow the example of Messrs. Knox and Gentz. But no man, we repeat, is or ought to be exempt at any period of his life from the liability to be captivated by female charms; and as the good Doctor was wanting in nothing that became a man, we may be sure that he fully appreciated Mrs. Thrale's delightful personality. Mr. Seeley in his pleasing biography of that lady correctly, if coldly, analyses the feeling with which Johnson regarded his hostess thus: " Perhaps he in some degree idealised the kind and sprightly dame who indulged his weaknesses, heaped his plate with dainties, forbore to count his cups of tea, and was able besides to cap his Latin quotations, and to join him in the critical remarks on English poets which he relished more than any other talk." No doubt this is true; but hardly, we think, the whole truth. Not, we firmly believe, that Johnson entertained for the lady a single feeling unworthy of his position. To have done so was not in the nature of that honest gentleman. But, from all we know of the man in his relations with the opposite sex, it is difficult for us to realise that his esteem for his hostess was merely compounded of a selfish regard for his own creature-comforts and a no less selfish appreciation of her ability to follow his remarks. The Doctor was made of other and better stuff than the cold-hearted egoist of the Seeleyan theory. We may be sure that he liked Mrs. Thrale for her own sake, and he would be most insensate and ungrateful not to have so liked her. As to loving her in the romantic sense, the character of the man and the nature of their relations alike preclude the notion of such a feeling at that time, whatever may have been the case afterwards, when the conditions under which they lived were altered. To him she was, then at least, in the light of a daughter; and he gave her freely the affection with which he was almost bound to requite her devotion to himself. If, when she became a widow and a middle-aged woman to boot; if, when the Divine Law and the discrepancy between youth and age no longer forbade the existence of a warmer sentiment on his part, he began to realise that his feeling towards her had all along been something beyond the interest natural to a guide, philosopher, and friend ; that she had become indispensable to him ; that he could not yield her up to another, what was there wonderful or derogatory in the circumstance ? We do not realise how strongly a plant is rooted in the soil until we attempt to pluck it up; and Johnson's love, from small beginnings, may have been growing imperceptibly during those sixteen years spent with the Thrales, so that he did not himself know how strong it was until the wrench of parting came. We do not say that this was actually the case; but it *might* have been; and if adopted as a hypothesis, it would serve to explain much that followed when, the household at Streatham being broken up, Johnson and the widow drifted apart.

The separation did not take place immediately on Mr. Thrale's death, and it certainly

Dr. Johnson as a Suitor. 57

was not of Johnson's seeking. On the contrary, he seems to have clung to the widow's skirts as long as he decently could. When, having found a tenant for the house at Streatham, Mrs. Thrale left for Brighton, Johnson left too; but in some mysterious way rejoined her on the road, when they travelled together the rest of the way and took up their quarters in the same boarding-house on arriving at that watering-place. This was in October, 1782; and early in the following year Boswell finds the Doctor staying at her house in Argyle Street, London. Meantime, however, their intimacy had not been of an unruffled character, and the widow was evidently growing tired of it—the fact being that her "guide, philosopher and friend" was very much in the way. Some time before her husband's death an interesting foreigner had been introduced at Streatham—a Signor Piozzi, who was young, personable and gifted, who sang delightfully to his own incomparable accompaniments, and altogether made himself very agreeable. During her husband's lifetime nothing save ordinary friendship existed between the two;

but upon being left a widow with some £3000 a year, Mrs. Thrale, whose first marriage had been one of convenience, entered into from deference to the wishes of parents on both sides, chose to think that, in making a second, she might please herself; and began to treat Piozzi after a fashion which showed that she would not object to him as a possible "future." That gentleman naturally responded to the advances of the rich and charming widow as became him; and before long their mutual attachment was no secret. How did Johnson comport himself whilst all this was going on? At first he does not seem to have entertained any suspicion on the subject. We read that, during the summer which Mrs. Thrale and her daughter spent at Streatham immediately after her bereavement, he came and went according to his wont. Miss Burney says of him at this time, "I really think he grows gayer and gayer daily, and more cheerful and pleasant": he dressed better, sometimes actually appearing "with silver buckles to his shoes"; and folk began even to couple his name archly with that of his fair hostess.

Boswell, in alluding to this period, observes: "Johnson's wishing to unite himself with this rich widow was much talked of; but I believe without foundation." "I believe so too," is written in the lady's hand opposite this passage; and possibly she was right. But when Mr. Piozzi's visits became more and more frequent, and the lady's manner to him more and more encouraging; when on the very first anniversary of her husband's death she discarded her weeds and quoted a list of possible suitors for her hand which was given in the newspapers, and in which the name of Piozzi figured, the Doctor began, as the saying is, to "smell a rat." He grew peevish, as he himself admits, and his visits to Streatham became fewer, whilst the time which they afterwards spent together in London was chequered by increasing coldness on the lady's side, and growing ill-humour on that of Johnson. A catastrophe was inevitable, and at last the blow fell. Johnson's temper having, as she alleges, become too much for her endurance, Mrs. Thrale fled to Bath. Their leave-taking, which proved a final one, was a great shock to the old man. In his diary (under 5th April, 1782) is the entry :—" I took leave of Mrs. T. I was much moved; I had some expostulations with her." She left for Bath the next day, and he writes—" I love Mrs. T. with a never-to-cease affection, and pity her more than I ever pitied any human being; and, if I did not blame, I could I believe almost die for her." Again on the 25th of April he sends this message to her—" Do not let Mr. Piozzi or anyone else put me out of your head, and do not think that anybody will love you like yours, &c." Did he, or did he not hope that, moved by these appeals, she would throw Piozzi over and take—no matter whom? There had been passages between them before which must have told her a good deal. In that pure melodious English which was natural to him when deeply moved, he had written to her while she was staying at Brighton in the autumn of 1877 : "Methinks you are now a great way off: and if I come, I have a great way to come to you; and then the sea is so cold, and the rooms are so chill! Yet do I love to hear the sea roar and my mistress talk .... I wish I was with you; but we are half the length of England asunder. It is frightful to think how much time must pass between writing this letter and receiving an answer." Did he fancy she would bear all this in mind? If so, the Doctor was a donkey, and thistles—not orange-blossoms— the fare that beseemed him. The widow's affections were by this time too deeply engaged for her to think seriously of breaking with Piozzi; but in deference to the remonstrances of her friends she made a show of doing so, and her lover returned to his native land. Her chagrin at the separation, however, could not be concealed, and she remained chafing at Bath, where her daughters had to adopt the unusual *rôle* of comforters to a love-sick mother. Meantime, she and Johnson corresponded; but not in the old delightful way. Her letters were constrained; his unsympathetic—nay, often harsh. The truth was that the old gentleman was in a very bad humour. He saw through the pretence of the Piozzi banishment, and hated the exile with all his heart—calling him " an ugly dog," when in fact, as Samuel Rogers, the poet, assures us, " Piozzi was a very handsome, gentlemanly and amiable person." Of the lady he said to Boswell, " Sir, she has done everything wrong since Thrale's bridle was off her neck"; and to Sir J. Hawkins—" Poor Thrale! I thought that even her virtues or her vices would have restrained her from such a marriage. She is now become a subject for her enemies to exult over, and for her friends, if she has any left, to forget or pity." But to herself came the passionate remonstrance found afterwards among her letters, "I have loved you with virtuous affection. I have honoured you with sincere esteem. Let not all our endearments be forgotten; but let me have in this great distress your pity and your prayers .... Do not, do not drive me from you, for I have not deserved either neglect or hatred." Silly old Doctor! It was not your letters she was longing for all the time, but those which bore an Italian postmark. At length she could endure the separation no longer; her health gave way, and Piozzi was summoned back. Soon after, a formal letter conveyed to Johnson

Dr. Johnson as a Suitor. 63

and her other guardians notice of her engagement. The intelligence seems to have deprived him of his senses for the time being: otherwise, he would scarcely have sent her the following outrageous and incoherent reply:—

"Madam,—If I interpret your letter right you are ignominiously married; if it is yet undone, let us once more talk together. If you have abandoned your children and your religion, God forgive your wickedness; if you have forfeited your fame and your country, may your folly do no further mischief. If the last act is yet to do, I who have loved you, esteemed you, reverenced you and served you, I who long thought you the first of womankind, entreat that, before your fate is irrecoverable, I may once more see you.

"I was, I once was, Madam,
"Most truly yours,
"July 2, 1784. SAM. JOHNSON."
"*I will come down if you permit it!*"

But the justly indignant lady did not permit it. She sent a withering, though not unmerited, reply, which concluded with the terrible sentence: "Until you have changed your opinion of Mr. Piozzi let us converse no more." This last blow crushed the spirit of the proud old man. He wrote her a penitent, pathetic letter—the swan-song of his affection —in which we find such expressions as " I breathe out one more sigh of tenderness, perhaps useless, but at least sincere." . . . "The tears stand in my eyes": takes a pitiful leave of her, and signs himself hers "with great affection." This letter bears a minute in the lady's hand : " I wrote him a very kind and affectionate farewell." It was all over. A fortnight afterwards she became the wife of Mr. Piozzi; and, before the year was out, Johnson slept in Westminster Abbey.*

Are we then to suppose that the two events are related as cause and effect? To do so would be mere assumption; for it must be remembered that Johnson was at this time a very old man, who had long been in a state of health that excited the apprehensions of his friends. Probably we have to deal with a coincidence for which the winter of 1784 is mainly responsible. Moreover, is it not upon

---

* The date of Mrs. Piozzi's marriage is 25th July, 1784; that of Dr. Johnson's death, 13th December, 1784.

record that, when Miss Burney saw him for the last time, a fortnight before his death, the following conversation took place :—

"I had seen Miss T. the day before.
"'So,' said he, 'did I.'
"I then said,—'Do you ever, Sir, hear from her mother?'
"'No,' cried he, 'nor write to her. I drive her quite from my mind. If I meet with one of her letters, I burn it instantly. I have burnt all I can find. I never speak of her, and I desire never to hear of her more. I drive her, as I said, wholly from my mind.'"

Whereupon his visitor found it expedient "wholly to change this discourse" by introducing the history of a Bristol milk-woman who had taken to literature and wrote "wonderfully." Was the dying man's outburst of passionate wrath anything more than the petulance of illness, or was it the despairing cry of a broken heart—who shall say? If the latter, then, indeed, my masters, some of us are in a parlous plight. No wisdom, no age, no infirmities, warrant us against the greatest misfortune that can befall a graybeard. Even in the quiet vale of years lurk pitfalls digged by mocking Love; and it behoves all of us who are treading the downward path so to order our steps that we be not taken therein, to our sore hurt and grievous shame.

Many of her own, and all Dr. Johnson's, friends were furious with Mrs. Piozzi because she married "that foreign scoundrel"; unjustifiably so, in the writer's opinion. Why should she not have married the man of her choice; and why should she not have chosen a proper man—not too young,* handsome, clever, accomplished, amiable, with nothing alleged to his disfavour, except that he was a foreigner and distinguished in one of the noblest of all arts? Her family were grown up and well provided for. She was rich, and could afford to please herself. Her first marriage had not been one of love, nor her husband by any means a model spouse. Over her married life had always hung a cloud, despite her efforts to brighten it—a cloud which in the end veiled her own attractions from her husband and, as we know, caused him to turn his eyes elsewhere. The fact was she

* He was just forty.

Mrs. PIOZZI.

had all through been more or less in disgrace. Expected to furnish the house of Thrale with an heir who would perpetuate its name and carry on its gigantic business, she had lamentably failed. Daughters came in plenty, but never a son that lived ;* and we may be pretty certain that, however her natural cheerfulness enabled her to conceal the fact, there were domestic heartburnings on this account. In other respects, too, her married life had not been without its worries. Thrale at one time lost his health, and got the business into serious financial difficulties, from which she had to extricate it herself—at no small sacrifice of health and ease. Now that she was emancipated from thraldom and anxiety, why should she not look forward to the enjoyment of life and love? She was only just turned forty; and are there to be no "cakes and ale," no "rocks and valleys" for rich and healthy widows in her plight? The idea is absurd; and equally absurd was the notion then prevalent that she would have done well, if bent

* We find her on one occasion recording her *thirteenth* disappointment.

upon changing her condition, to select Dr. Johnson as her partner. We may admire and love the Doctor as much as we like and he deserves; but when the choice lay between him and the handsome, gentlemanly Piozzi, as a husband, we need not wonder at the Italian having carried the day.

Whether, in the history of her relations with Dr. Johnson prior to her second marriage, the lady's conduct be altogether free from blame is a more difficult question. One of Her Majesty's counsel learned in the law[*] holds that it was; and there is a weight of opinion behind him. Others, on the contrary, maintain that she knowingly and capriciously tried to captivate the venerable sage, and having led him on to an infatuation unbeseeming his grey hairs, behaved pretty much as Vivien did to Merlin. I profess that the decision of this knotty point is quite beyond my humble powers; the male intelligence is unfitted to cope with its refinements. But you, madam, who chance to honour these pages with your regards—what do *you* think?

[*] The late Mr. Hayward, Q.C.

Dr. Johnson as a Suitor. 69

Here ends the tally of Johnson's loves: for although in at least one other case, that of Miss Boothby, there are circumstances which go to prove the existence of a tenderer feeling than mere friendship, the weight of evidence is not sufficient to warrant a definite conclusion to that effect. The list is, we venture to think, a very fair one, considering the Doctor's personality, the dignity of his character, and the nature of his pursuits. In fact, few ancients have left so brave a show; but then, Johnson was a very exceptional man who, save for the mishap which befel him in that Piozzi affair, might have gone on loving until he had established a "record."

III.

DR. JOHNSON AS A MAN OF FASHION.

THE popular notion of a great man is always incomplete. It is drawn only from that side of his character most frequently presented, and fails to take his other aspects into account. Johnson seated in his arm-chair at the Mitre, calling out "Who's for poonsh?" Johnson rolling down Fleet Street, touching the posts as he goes along; Johnson at the club, heckling, and brow-beating, and laying down the law; Johnson in his chambers, balancing himself on a three-legged chair, and singeing the top of his wig in the candle as he is engaged in tearing the heart out of some book; Johnson shabby and unkempt; rough to look at and dangerous to approach, is the Johnson of nine men out of every ten who think about him at all. Lord Macaulay has drawn the picture for us in his own masterly way—"In the foreground is that strange

Dr. Johnson as a Man of Fashion. 71

figure which is as familiar to us as the figures of those among whom we have been brought up—the gigantic body, the huge massy face, seamed with the scars of disease, the brown coat, the black worsted stockings, the gray wig with the scorched foretop, the dirty hands, the nails bitten and pared to the quick. We see the eyes and mouth moving with convulsive twitches; we see the heavy form rolling; we hear it puffing; and then comes the 'Why, sir!' and the 'What then, sir?' and the 'No, sir!' and the 'You don't see your way through the question, sir!'" All true enough; but still, not the whole truth.

The fact is that Johnson led two distinct lives: one, the most familiar to us, among his own set—the Burkes, Reynoldses, Garricks, Langtons, Goldsmiths, Beauclerks, &c.; the other, as a member of society, a leading figure in the fashionable world of his day. And we must recollect that the fashionable world was, at that time, distinctly literary in its character. The leaders of *ton*, the great ladies who then set the fashion, had, for purely social reasons at first, brought this state of things about.

The vice of card-playing at fashionable assemblies had increased to a degree which threatened social disaster. Men lost their fortunes, and women their reputations, over whist, ombre, or quadrille, at card-tables spread in the most select drawing-rooms, and the good name of English society was fast falling into disrepute. Then a few noble-minded women cast about for some means of arresting this downward tendency; and that which presented itself to them was to interdict card-playing at their own parties, at the same time providing a substitute in the presence of clever and distinguished people who, by giving a brilliant tone to the conversation, might render it sufficiently attractive to reconcile their guests to the deprivation. The leaders of this movement were so influential that it was bound to succeed at first. Even the most inveterate card-players could not afford to absent themselves from the reunions of such very grand dames; and, by-and-by, as is the nature of society, that which at first seemed ridiculous ended by becoming the fashion. Every woman of any social preten-

Dr. Johnson as a Man of Fashion. 73

sion was fired with ambition to be thought literary; until, according to the usual fate of such movements, the thing was pushed to an extreme which almost defeated its own ends. Gradually, what came to be known as "The Blue-stocking Society," was evolved from this new agitation; and although destined to end in a *fiasco*, it lasted long enough, and made its influence enough felt, to deserve more than a passing mention. The name originated, it is said, in the reunions held by such prominent leaders as Mrs. Montagu, Mrs. Vesey, Mrs. Ord, &c., at which, in order to encourage gentlemen to attend, certain dispensations as regards the use of evening dress—then a very serious affair—were granted. Among the visitors who availed themselves of that privilege was a Mr. Benjamin Stillingfleet, who actually appeared in hodden-gray, or blue, worsted stockings, instead of the black silk hose usual on such occasions. This peculiarity so impressed another visitor, Admiral Boscawen, who did not half like such easy-going ways, that he declared they had better call themselves "The Blue-stocking Society," meaning thereby "a set of people who did not care how they went," and the ladies who composed it "Blue Stockingers." The name took hold at once, and was afterwards corrupted into "Blue Stockings," and eventually into "Blues."* Hannah More helped to fix it by her poem, 'Bas Bleu,' in which many of the most conspicuous Blue-stockings are celebrated; but it is rather depressing to learn that Mr. Stillingfleet, its founder, proved a traitor to the cause. In one of Mrs. Montagu's letters he is upbraided with having "left off his old friends and his blue stockings." Of course, as Boswell informs us, "Johnson was prevailed with to come sometimes into these circles, and did not think himself too grave even for the lively Miss Monckton (now Countess

---

* The above is the most generally received account of its origin; but there are many others, as for instance that of Lady Crewe, quoted by Hayward, which traces it to a Madame de Polignac having presented herself at Mrs. Montagu's in blue stockings, then the rage in Paris, which attracted the ladies present so much that they lost no time in copying her. Madame d'Arblay employs the original terms in her Diary (Vol. i. p. 326): "Who would not be a Blue-stockinger at this rate?"

Dr. Johnson as a Man of Fashion. 75

of Corke) who used to have the finest *bit of Blue* at the house of her mother, Lady Galway." The movement spread outside London, and as it overflowed the provinces became more shallow until at last it began to excite contempt. For, departing from the intention of its founders, those eager missionaries who promulgated its doctrines in remoter parts introduced divers abuses. Thus, as will appear later on, at her villa near Bath, Lady Miller practised certain strange rites. She set up a large stone vase called "the Tully Vase," fetched all the way from Rome, and which now stands in the Victoria Park; and in this vase were deposited on certain festive occasions the votive offerings of her modish guests, consisting of *bouts-rimés* and other rhymed vanities. At such times a fair priestess, attired in sacerdotal garments, would draw forth from the vase, which was decked for the nonce with ribbons and garlands, the various metrical inanities, and adjudge the prize, consisting of a myrtle wreath or other vegetable contrivance, to the writer of what offering soever was deemed most worthy. It is hardly necessary to say that Johnson objected to the degradation of a movement to which he had lent his support. "I wonder," he growled, "how people were persuaded to write in that manner for this lady." BOSWELL: "I named a gentleman of his acquaintance who wrote for the vase." JOHNSON: "He was a blockhead for his pains." BOSWELL: "The Duchess of Northumberland wrote."* JOHNSON: "Sir, the Duchess of Northumberland may do what she pleases; nobody will say anything to a lady of her rank"—which is profoundly true. But however much he might disapprove of its vagaries, it was not easy for the Doctor to withdraw himself from all connection with a movement which had such distinguished leaders. The charm of good society is a charm which grows more powerful with use; nor, having once been experienced, can its use be abandoned without a pang. Johnson was an aristocrat at heart; he preferred the society of those above him in the social scale, though not the man

* It is interesting to know that the subject of Her Grace's muse was "A buttered muffin."

to separate himself from those of his own level. "High people, sir, are the best," was one of his favourite maxims, which he rarely failed to carry into practice when he had the chance. Besides, being, as we have shown, extremely susceptible to the glamour of female companionship, he was not disposed to sever his connection with those who were able to afford it to him in its most attractive form. For we must recollect that society had identified itself with the movement; also, that it was the fashion to be intellectual. And here we may point to the distinction between the "New Woman" of the eighteenth century and her successor of our own times. However "advanced" the former might be, she never sought to divest herself of her femininity. The notion of the divided skirt had not dawned upon her; the problem of sexuality troubled her not. If it had been suggested to her that she should engage in athletic pursuits of the same nature as those practised by men, she would probably have recoiled from the idea with horror, at the same time pointing out that history showed how such a departure from womanly traditions had heralded the decline of the Roman Empire. She might dabble in republicanism, denounce the injustice of the laws affecting her sex, protest against the excessive domination of men, and assert her right to education, representation, and all the rest; but she was content to remain, a woman, in a sphere absolutely distinct from that of man. Hence, we do not find that the male sex resented the Bluestocking agitation, as, indeed, why should they? The Blues were just as feminine as other women, with the superadded charm of higher culture. Men like Johnson not only fancied a pretty face and graceful ways, but wanted something in addition—namely, as he himself put it, that the owner of such attractions "should be capable of understanding him and of adding something to the conversation." All this was to be found in the New Woman of that day; and we can therefore see good reason why our worthy Doctor was loth to quit her society. At all events, he continued to attend the conversaziones of Mrs. Montagu and her compeers; and, as he

Dr. Johnson as a Man of Fashion. 79

prided himself upon being "a very polite man," we may be sure that, whatever society he happened to be in, he "comported himself accordingly." The scorched wig, the frayed shirt, the baggy old snuff-coloured suit, were discarded on such occasions together with the "bow-wow" manner, and he appeared like what he really was, when he chose, a refined gentleman. How glad these fashionable dames were to have him: how they petted and flattered him, has already been told; but it may be as well to give an extract from one of Langton's letters, in which he describes an evening spent at Mrs. Vesey's :—

"The company consisted chiefly of ladies, among whom were the Duchess of Portland, the Duchess of Beaufort, whom I suppose, from her rank, I must name before her mother, Mrs. Boscawen, and her elder sister Mrs. Lawson, who was likewise there; Lady Lucan, Lady Clermont, and others of note both for their station and understanding. Among the gentlemen were Lord Althorpe, Lord Macartney, Sir J. Reynolds, Lord Lucan, Mr. Wraxall, Dr. Warren, Mr. Pepys, the Master in Chancery, whom I suppose you know, and Dr. Barnard, the Provost of Eton. As soon as Dr.

Johnson was come in and had taken a chair the company began to collect round him till they became not less than four, if not five, deep; those behind standing and listening over the heads of those that were sitting near him."

Nor was it in such assemblies alone that he was made welcome. Wherever witty or clever people were to be found Johnson had the *entrée*. To him the green-rooms of the theatres were gladly open; nor was he slow to avail himself of the privilege. In fact, many of his most valued friends were among the distinguished actresses of the day. To give a complete list of his female acquaintance would occupy too much space; but a few of the most prominent may be selected—those whom he particularly liked, and those whom, it is to be feared, he regarded with other feelings. For convenience sake, they shall be briefly described here in that order which, as we know on good authority, is better than no order at all, and with the omission of unnecessary detail.

There was first of all "The Abington."

Now The Abington, though she had many traducers, was a great power in the land. She mixed with the very best people, who were, at least, glad to know her, even if they pulled her reputation to pieces among themselves. Though with good blood in her veins, she had been under a cloud, and began life as " Fanny Barton "—afterwards to be known as " Nosegay Fan," the flower-girl. But the artistic bent was not long in showing itself. How she picked up a few accomplishments, Heaven only knows; but after a little while flower-hawking was given up, and she is found singing and reciting at taverns and coffee-houses in and about Covent Garden. Next, it somehow dawned upon her that she needed polish; and the only way of obtaining the latter which then presented itself was to enter the service of a French milliner, from whom she acquired two things most valuable to an actress, viz., taste in dress and a knowledge of French. Her progress in education now became rapid. She made herself acquainted with the best French authors; and not only learned to speak and write French correctly, but also to converse in Italian. At the age of eighteen she made her first appearance on the stage, and at once achieved a fair amount of success. Four years afterwards she is described in the bills as "Mrs. Abington," having in the meanwhile married her music-master. The marriage did not prove a happy one, and she eventually separated from her husband—to whom, however, she made a yearly allowance on the condition that he should keep away from her. Though by this time an established favourite at Drury Lane, she found that Mrs. Pritchard and Mrs. Clive stood in the way of her advancement to the very highest rank in her profession, and determined to seek fortune elsewhere. Having accepted an engagement in Dublin, she took that metropolis by storm—not only her acting, but her exquisite taste in dress attracting her audiences, among whom "The Abington Cap" became a watchword. The fame of her success reached Garrick, who thereupon implored her to return, and for the next eighteen years she filled leading parts at Drury Lane. It is curious to note that when

Mrs. ABINGTON.

forty years of age she was the first "Lady Teazle," and looked the youthful part as well as she acted it ; while, at sixty, a well-known critic declares of her that "she can still give to Shakespeare's "Beatrice" what no other actress in my time ever conceived." In private life she was most charming, and soon attained a distinguished social position by her beauty, wit, and cleverness—aided by her exquisite taste in dress. Her features were not strictly regular ; but her expression amply compensated for that drawback, and she possessed a singularly elegant figure. If her voice was not particularly good, her articulation was wonderfully distinct and harmonious. In a word, she was, and deserved to be, "the fashion" ; and Boswell admits that, on this account, Johnson was highly flattered by her attentions, and "loved to bring forward his having been in the gay circles of life." Indeed, she was splendidly attentive to the old man. We learn from Baretti how she invited him to dinner, " and took pains to distinguish him above all her guests, who were all people of the first distinction " ; and we further read in Boswell how, "having supped with Mrs. Abington in the company of certain persons of fashion, he (Johnson) was much pleased with having made one in so elegant a circle." He repaid her kindness in the way most flattering to an artist, by attending her performances, though unable to see or hear what was going on. One day, Boswell states, "he told us that he was engaged to go that evening to Mrs. A.—'She was visiting some ladies where I was visiting, and begged that I would come to her benefit. I told her I could not hear, but she insisted so much on my coming that it would have been brutal to refuse her.'" Afterwards one of the company was foolish enough to ask, "Why, Sir, did you go to Mrs. A.'s benefit? Did you see?" JOHNSON: "No, Sir." "Did you hear?" JOHNSON: "No, Sir." "Why, then, Sir, did you go?" JOHNSON: "Because, Sir, she is a favourite of the public; and when the public cares the thousandth part for you that it does for her I will go to your benefit, too." The Abington had yet another attraction for the Doctor—she gave uncommonly good dinners; and we find him once telling Mrs. Thrale at her own table that Mrs. Abington's jelly was superior to hers. Such a combination of perfection was irresistible; and to the venerable sage she remained an object of worship all his life.

Of Miss Adams we know little, except that she was the daughter of Dr. Adams, Master of Pembroke College, Oxford; one of Johnson's prime favourites, and the heroine of that little episode narrated at page 20, in which a coffee-pot played a distinguished part. But we *do* know that "this lady's good qualities, merit, and accomplishments, and her constant attention to Dr. Johnson, were not lost upon him." Of course, he teazed her sometimes; but young ladies do not object to a little banter from old gentlemen of the right sort. For instance—

"Miss Adams mentioned a gentleman of licentious character, and said, 'Suppose I had a mind to marry that gentleman, would my parents consent?' JOHNSON: 'Yes, they'd consent, and you'd go. You'd go, though they did not consent?'

MISS ADAMS: 'Perhaps their opposing might make me go?' JOHNSON: 'O, very well; you'd take one whom you think a bad man, to have the pleasure of vexing your parents. You put me in mind of Dr. Barrowby, the physician, who was very fond of swine's flesh. One day, when he was eating it, he said, "I wish I was a Jew,"—"Why so? (said somebody); the Jews are not allowed to eat your favourite meat." "Because (said he) I should then have the gust of eating it, with the pleasure of sinning."'"

"Miss Adams," continues Boswell, "soon afterwards made an observation which pleased him much: he said, with a good-humoured smile, 'That there should be so much excellence united with so much depravity is strange.'" Johnson used to declare that he was never so happy as when staying at her father's house.

Of Miss Adey we know still less. She was the niece of Mrs. Cobb, a widow lady, who lived in an agreeable sequestered place called "the Friary," close by Lichfield, where we find Johnson and Boswell breakfasting with the two ladies, and the former behaving to them "with a kindness and easy pleasantry such as we see between old and intimate acquaintances." Boswell paid an unconscionably early visit to the Friary a year afterwards on his own account, and wrote how his visit "at first occasioned some tumult in the ladies, who were not prepared to receive *company* so early, but my *name*, which has by wonderful felicity come to be so closely associated with yours, soon made all easy." Both aunt and niece were in fact "great admirers of Dr. Johnson," and would gladly have suffered a greater nuisance than Boswell's untimeous visit for his dear sake; but it was hard on them to be surprised thus *en déshabille*, and in company with their unbidden guest to resume their seats at the breakfast-table "which they had quitted with some precipitation."

It has already been stated that "Mrs. Boothby" occupies a doubtful position in the list of Johnsonian belles. Many circumstances would incline us to believe that she was something more to him than a "friend"; but as his conduct is, in her case, marked by inconsistencies which do not present themselves in his acknowledged love affairs, we have not included her in the catalogue of his "loves." Miss Hill Boothby, the daughter of a country baronet, made Johnson's acquaintance through being a friend of that Mrs. Fitzherbert whose intellect he so highly praised.* He soon began to "esteem" her; though he told Mrs. Thrale that "she pushed her piety to bigotry, her devotion to enthusiasm; that she somewhat disqualified herself for the duties of this life by her perpetual aspirations after the next." Such was, however, the purity of her mind, he said, and such the graces of her manner, that Lord Lyttelton and he used to strive for her preference with an emulation that occasioned hourly disgust and ended in lasting animosity. Here, then, we have in all probability the secret of Johnson's devotion—he wanted to cut out "that fellow Lyttelton"; and so far did he carry this jealousy as to insert a biography of that nobleman in his 'Lives of the Poets,' in which himself and his works are pretty roughly handled. Meantime, he continued to pay certain half-hearted attentions to the cause

* *Vide*, p. 29.

Dr. Johnson as a Man of Fashion. 89

of so much bad blood. The lady used to suffer from indigestion, and he recommended her as a remedy "dried orange-peel finely powdered taken in a glass of hot red port"; but there is not very much in that. He also wrote letters to her; but they are constrained, have not the true ring and, to tell the truth, are rather stupid. We give one by way of example :—

"January, 1755.

"Dearest Madam,—Though I am afraid your illness leaves you little leisure for the reception of any civilities, yet I cannot forbear to pay you my congratulations in the new year, and to declare my wishes that your years to come may be many and happy. In this wish, indeed, I include myself who have none but you on whom my heart reposes (*O, Doctor, Doctor !*); yet surely I wish you good, even though your situation were such as should permit you to communicate no gratification to, dearest, dearest Madam,

"Yours, etc.,
"SAM. JOHNSON."

Did anyone ever read a more lumbering *billet doux*? One can see there is no heart in it, though he assures her that she is the only being on whom that organ "reposes"; and can feel that were Lord Lyttelton out of the way Johnson's ardour would be a very lukewarm affair. But the rivalry serves to keep him hanging on my lady's skirts. "You may see," said he to Mr. Thrale when 'The Lives of the Poets' appeared, "that dear Boothby is at my heart still. She *would* delight in that fellow Lyttelton's company though, all that I could do; and I cannot forgive even his memory the preference given by a mind like hers." Altogether the Doctor does not appear to much advantage in the matter. Miss B. evidently had something like a tenderness for his lordship; and it was Johnson's vanity which probably served to keep two worthy people from being made happy, whilst his own affections were not the least involved. Yet we cannot be quite certain on this point, for Baretti told the Thrales that "when this lady died, Dr. Johnson was almost distracted, and the friends about him had much ado to calm the violence of his emotion." *

The name of Mrs. Bosville, of Gunthwait, in Yorkshire, occurs but once in Boswell's 'Life of

* 'Piozzi Anecdotes.'

Johnson'; but, judging from the context, she may have been something more than a chance acquaintance. Having met them on one occasion at the Pantheon, we are informed that she entered into conversation with Johnson and his satellite, and that the Doctor afterwards remarked to the latter—"Sir, this is a mighty intelligent lady." Perhaps Boswell's reason for dwelling upon the interview is that, just about the time it took place, he was debating whether it would not be advisable for him to pay his addresses to her daughter, a young lady with excellent prospects. But the project fell through, and the wife of "Squire Godfrey Bosville" disappears from his pages. One would like to know more of this "mighty intelligent lady."

The Honourable Mrs. Boscawen, wife of Admiral Boscawen, was not only prominent in Blue-stocking society, but a person of the very highest *ton;* and the "assemblies" at her house in St. James's Square were far more exclusive than those of her literary rivals. In addition she was a very charming woman, endowed with so many gifts as to be styled, in *Blue* jargon, "The accomplished Mrs. Boscawen." One of the numerous bards who hymned the praises of eminent Blue-stockings described her thus in eloquent numbers :—

"Each art of conversation knowing,
High bred, elegant Boscawen."

Which tribute Miss Burney confirms by declaring her to be "all elegance and good-breeding," and Boswell, with that awe which he reserves for great people, affirms, "If it be not presumption in me to praise her, I would say that her manners are the most agreeable, and her conversation the best, of any lady with whom I ever had the happiness to be acquainted." But it is to Hannah More that we are indebted for glimpses of that inner life to which the great lady's particular friends alone were admitted. Writing to one of her sisters, of a Sunday night, she says :—"I have been at Mrs. Boscawen's. Mrs. Montagu, Mrs. Carter, Mrs. Chapone, and myself only were admitted. We spent the time, not as wits, but as reasonable creatures; better characters, I trow. The conversation was sprightly, but serious. I have not enjoyed an afternoon so much since I have been in town."

But we are sorry to learn from her that "Mrs. Boscawen's life has been a continued series of afflictions, which may almost bear a parallel with those of the righteous man of Uz." One hopes the Admiral was not responsible for this state of things. Again, "On Tuesday Mrs. Boscawen carried me to Glanville; we had the pleasantest *tête-à-tête* day imaginable, and walked about and sat under the spreading oaks, and ate our cold chicken, and drank our tea, as happy folks are wont to do." From all which we may gather that Mrs. Boscawen was distinctly what is known as "a woman's woman," though in her capacity of a Blue-stocking she was necessarily brought much into contact with the rougher sex. One great friend she had amongst the latter—he who was everybody's friend, Horace Walpole, whose letters abound with references to her. But there is nothing to show that she was ever on terms of very ardent intimacy with Dr. Johnson, whom she may have thought too loud and disputatious for those quiet little *réunions* in which her soul delighted. They met frequently, as was inevitable, in the outside world, and each valued the good qualities of the other; but, so far as we know, that was all.

Though a reconciliation eventually took place between Dr. Johnson and his quondam hostess, Mrs. Boswell, the Doctor remained so long on distant terms with the lady in question that she may, we fear, be counted among his "aversions." The feud between them dated from his visit to Boswell in 1771; for the first allusion to it occurs in a letter announcing his return to London, which contains the ominous words: "I know Mrs. Boswell wished me well to go; her wishes have not been disappointed."

"In this," remarks Boswell, "he showed a very acute penetration. My wife paid him the most assiduous and respectful attention while he was our guest, so that I wonder how he discovered her wishing his departure. The truth is that his irregular hours and uncouth habits, such as turning the candles with their heads downwards, when they did not burn bright enough, and letting the wax drop upon the carpet, could not but be disagreeable to a lady. Besides, she had not that high admiration of him which was felt by most of those who knew him; and, what was very natural to a female mind, she thought he had too great influence over her husband. She once in a little warmth made, with more point than justice, this remark upon that subject: 'I have seen many a bear led by a man; but I never before saw a man led by a bear.'"

No doubt any lady with Scotch housewifely instincts must have been sorely tried by the Doctor's ways, and her jealousy on account of her husband's infatuation was natural enough; but she must have rather forgotten her duties as a hostess to have raised misgivings in the bosom of her self-complacent guest. At all events, she managed to inflict a wound which long rankled in the Doctor's memory, and probably was never thoroughly healed by the balsam which poured into it freely enough when he was out of the way. He is ever recurring to the subject in his letters to Boswell—"Make my compliments to Mrs. B. and tell her that I do not love her the less for wishing me away." "Pray teach Veronica (their daughter) to love me. Bid her not mind mamma." "She (Mrs. B.) knows that she does not care what becomes of me." "I know that she (Mrs. B.) does not love me; but I intend to persist in wishing her well till I get the better of her." "My compliments to Mrs. B. if she is in a good humour with me." "I wish you, my dearest friend, and your haughty lady (for I know she does not love me), and the young ladies, and the young Laird, all happiness. Teach the young gentleman, in spite of his mamma, to think and speak well of me," &c. &c. &c. If the lady had erred, her punishment was terrible, and the Doctor's vindictiveness does not show him at his best. At last, whether from compunction or weariness, Mrs. Boswell struck her flag, and the shape which her overtures of peace took was eminently judicious. First of all she wrote him a kind letter, which was comparatively ineffective; for his answer concludes thus, after the usual civilities: "Pray take care of him (Boswell) and tame him. The only thing in which I have the honour to agree with you is, in loving him; and while we are so much of a mind in a matter of so much importance, our other quarrels will, I hope, produce no great bitterness." Slightly discouraged, perhaps, by this response, the lady made her next advances through her husband, who writes: "She (Mrs. B.) begs you may accept of her best compliments. She is to send you some marmalade of oranges of her own making." This evidently mollified the "bear" (who had a sweet tooth); for the response, though a growl, contains a note of kindness:—"Tell Mrs. Boswell that I shall take her marmalade cautiously at first. *Timeo Danaos et dona ferentes.* Beware, says the Italian proverb, of a reconciled enemy. But when I find that it does me no harm, I shall then receive it and be thankful for it, as a pledge of firm, and I hope unalterable, kindness. She is, after all, a dear, dear lady." At last, the marmalade being received and comfortably disposed of, the "bear" proceeded to climb down as follows:—

DR. JOHNSON TO MRS. BOSWELL.

" Madam,—Though I am well enough pleased with the taste of sweetmeats, very little of the pleasure which I received at the arrival of your jar of marmalade arose from eating it. (*Qy.*) I received it as a token of friendship, as a proof of reconciliation, things much sweeter than sweetmeats; and upon this consideration I return you, dear Madam, my sincerest thanks. By having your kindness I think I have a double security for the continuance of Mr. Boswell's, which it is not to be expected that any man can long keep when the influence of a lady so highly and so justly valued operates against him. Mr. Boswell will tell you that I was always faithful to your interest, and always endeavoured to exalt you in his estimation. You must now do the same for me. We must all help one another, and you must now consider me as, dear Madam,

"Your most obliged and most humble servant,
"SAM. JOHNSON."

And so, the feud was stayed.

Not to be familiar with the general history of Miss Fanny Burney, afterwards Madame d'Arblay, is to argue oneself a philistine ; and we shall, therefore, confine our notice of that lady's adventures and achievements to the bounds of her intimacy with Dr. Johnson. This may be said to have commenced with the publication of 'Evelina,' the comic parts of which took his fancy when read by her father at Mr. Thrale's. After reading the

Dr. Johnson as a Man of Fashion. 99

book himself, he declared "there were passages in it which might do honour to Richardson," and that he "could not get rid of the rogue." She was then asked by the Thrales to meet him at dinner, when she was placed next him, upon which, with his usual gallantry, he remarked: "Sitting by Miss Burney makes me very proud to-day"; and Mrs. Thrale thought it best to warn her: "You must take great care of your heart if Dr. Johnson attacks it; for I assure you he is not often successless." "What's that you say, Madam?" cried he; "are you making mischief between me and the young lady already?", and forthwith drank her health. In fact, she had gained his heart; and, as it was never the Doctor's way to mask his preferences or dislikes, he from that day began to praise, encourage, and even pit the girl-author against rivals of her own sex, such as the celebrated Mrs. Montagu. "Down with her, Burney!—down with her!—spare her not!" he once cried. "Attack her, fight her, and down with her at once! You are a rising wit, and she is at the top; and when I was beginning the world, and was nothing and nobody, the joy of my life was to fire at all the established wits, and then everybody loved to halloo me on . . . . So at her, Burney—at her, and down with her!" His pet name for her was "little Burney." He gave her Latin lessons, and was very good-humoured over them, too. He praised her to her face and behind her back. Boswell once happening to mention 'Cecilia' in the course of a rather flagging conversation, Johnson (with an air of animated satisfaction) exclaimed, " Sir, if you talk of ' Cecilia,' talk on!" He went further ; for we are told in the 'Diary' how, upon her once meeting him at Brighton, " Dr. Johnson received me with his usual goodness, and with a salute so loud, that the two young beaux, Cotton and Swinerton, have never done laughing about it." When, the old man being very unwell, she went to see him at Bolt Court, he received her with great kindness, and bade her come oftener ; and when " a queer man of a parson," who was also present on the occasion, remarked that Dr. Johnson had " made him read 'Cecilia,'" the invalid broke out with—

FANNY BURNEY.

"*Made* you, Sir? You give an ill account of your own taste or understanding if you wanted any *making* to read such a book as 'Cecilia.'" At a subsequent visit he asked after her brother Charles, and said: "I shall be glad to see him; pray tell him to call upon me;" and upon her thanking him for the permission, "I should be glad," said he, still more kindly, "to see him if he were not your brother; but were he a dog, a cat, a rat, a frog, and belonged to you, I must needs be glad to see him." We shall give her account of another visit in her own words:—

"The dear Doctor received me with open arms.

"'Ah, dearest of all ladies!' he cried, and made me sit in his best chair.

"He had not breakfasted.

"'Do you forgive my coming so soon?' said I.

"'I cannot forgive you not coming sooner,' he answered.

"I asked if I should make his breakfast, which I have not done since we left Streatham; he readily consented.

"'But, Sir,' quoth I, 'I am in the wrong chair.' For I was away from the table.

"'It is so difficult,' said he, 'for anything to be

wrong that belongs to you, that it can only be I am in the wrong chair, to keep you from the right one.' "And then we changed."

It is pleasant to read of the friendship between the two; it lasted while the Doctor lived, and was highly creditable to both parties. After the old man's death she did not forget his almost dying injunction, to "stand by him and support him, and not hear him abused when he was no more and could not defend himself." When Boswell, with characteristic bad taste, once began to tell stories of his old friend, acting them with incessant buffoonery, Miss Burney stood up for the dead lion. "I told him frankly that if he turned him into ridicule by caricature, I should fly the premises," and the narrator was obliged to modify his humour. She never fancied Mr. Boswell afterwards; and tells us how, when King George the Third once expressed to her his despair and disappointment at not finding more about her in 'The Life,' "I ventured to assure him how much I had myself been rejoiced at this very circumstance, and with what satisfaction I had reflected upon having very seldom met Mr. Boswell, as I knew there was no other security against all manner of risks in his relations." She was seldom so bitter as this; but to be ignored in that horrid way would have provoked a saint.

It is likely that Mrs. Elizabeth Carter was one of the most learned women of her day, and it is certain that she was one of the best—good, amiable, conscientious, and thorough. Born in 1717 at Deal, where her father held a perpetual curacy, she lost her mother at the age of ten; and from that time her father undertook her education. A good classic and Hebrew scholar himself, he thought that the best thing he could do for his daughter was to make her the same; and began with Greek and Latin. But her tender mind was yet unequal to the strain he placed on it; she seemed to have a natural incapacity for dead languages, and her progress was so slow that her father in despair proposed to give up his attempt. It was then that the true nature of the girl revealed itself. Sooner than disappoint him, she resolved to compass his object at any cost; and by immolating herself on the altar of duty, succeeded. As the management of the little household devolved upon her, she had not sufficient time during the day for the accomplishment of her purpose, and her nights were made to suffer in consequence. By such artifices as wet towels round her head, taking snuff, and chewing green tea, she managed to study far into the night; and she further contrived an unholy machine, consisting of a great weight suspended by a string which passed near her chamber-light in such a way that the flame was sure to reach it by a certain hour in the morning, when down would come the weight, and up would spring the wretched girl all unrefreshed, to grapple with her classic foes. Of course, she paid the penalty which Nature imposes for abuse of her powers; and the overwrought brain had its revenge in the shape of severe headaches, from which she suffered all through her future life. But her end was achieved. Not only did she master Greek and Latin, but developed a wonderful capacity for the study of languages in general. She became proficient in Hebrew, French, Italian, Spanish, and German; and later on, taught herself Portuguese and Arabic. Then she dabbled in astronomy, dived deep into ancient and modern history, and floundered considerably in ancient geography. Nor were pursuits more native to her sex forgotten; she was an expert needlewoman and housewife, a good cook, and a skilful performer, not only on the spinet, but on that oddly-chosen instrument, "the German flute." Truly, an admirable woman!

About the age of seventeen this female Crichton entered the world of letters by the usual portal; she composed verses, some of which, signed "Eliza," found their way into 'The Gentleman's Magazine,' whereof Cave, her father's friend, was the publisher. Cave introduced her to Johnson, his editor, who, delighted by her classical attainments, gave her every encouragement, paid her the suitable compliment of a Greek epigram which appeared in the pages of the magazine, and declared that she "ought to be celebrated in as many languages as Lewis le Grand";

indeed, his friendship with her lasted all through the rest of his life, a period of nearly fifty years. She now busied herself with translations, and published one of De Croussaz's 'Examen,' of Pope's 'Essay on Man,' which was actually attributed to Johnson and proved a great success. Sundry translations of scientific treatises succeeded, such as Algerotti's 'Newton for Young Ladies,' though of these she was rather shy in after life; but her crowning feat of this kind was a translation of Epictetus, with introduction and notes—a valuable and successful work. Meanwhile she was an occasional contributor to the *Rambler*,* and did her best to advance the interests of that journal, which had become unpopular owing to Johnson's introduction of so many "hard words," with the view, as people hinted, of compelling readers to buy his dictionary. "Many a battle for him I have fought in the country, but with little success," she owned afterwards. 'Poems on Several Occasions' followed, which ran through several editions, but which modern readers would not

* Nos. 44 and 100 are by her.

ELIZABETH CARTER.

greatly relish. At last her reputation for learning and authorship became so great that the Government bestowed on her a yearly pension of £150, which she enjoyed until she had attained the great age of eighty-eight; and though her closing years were marked by physical infirmity, her mind remained clear to the last. She never married—probably never had time to do so, though not destitute of personal attractions. Miss Burney, who did not make her acquaintance till she was comparatively aged, says: "She is really a noble-looking woman; I never saw age so graceful in the female sex yet; her whole face seems to beam with goodness, piety, and philanthropy." Indeed, she was a thoroughly good, sensible, and amiable woman, who knew everybody worth knowing, and was liked by all. Johnson had a warm affection for her. Writing to her in 1756 about his publisher's death, he says: "Poor dear Cave! I owed him much; for to him I owe that I have known you." And when another lady was being praised in his hearing on account of her learning, he broke in with, "A man is in general better pleased when he has a good dinner upon his table than when his wife talks Greek. My old friend, Mrs. Carter, could make a pudding as well as translate Epictetus." On another occasion he remarked to Boswell : "I dined yesterday at Mrs. Garrick's with Mrs. Carter, Miss Hannah More, and Miss Fanny Burney. Three such women are not to be found." * When discussing the merits of a contemporary, he observed that B. understood Greek better than anyone he had ever known "except Elizabeth Carter." Indeed, it is upon her knowledge of Greek that her reputation is chiefly founded ; for while her poems have sunk into deserved oblivion, her translation of Epictetus still maintains her reputation for sound scholarship—solidity, not brilliancy, being what chiefly distinguished this gifted lady among her contemporaries. Wraxall terms her "the Madame Dacier of England," "the only woman who was qualified to meet Johnson on equal terms" ; and

---

* It should be mentioned that he added, "except Mrs. Lenox, who is superior to them all"; but at this time he was *raffolé* of the fair Charlotte.

with all this, she remained so thoroughly feminine, pious, modest, sensible, and kind, as to redeem the character of the Blue-stocking sisterhood from much of the odium which the vagaries of other ladies had brought upon it.

Mrs. Cholmondeley was a being of a vastly different order, whom Johnson describes as "a very airy lady." The brilliant wife of the Honourable and Reverend Robert Cholmondeley, and sister of the celebrated "Peg Woffington," did not found her claims to social distinction upon her intellectual attainments, though of some repute as a wit and even as a critic. Miss Burney brackets her with Mrs. Thrale as being "severe and knowing"; and adds, "she has shown so much penetration and sound sense of late that I think she will bring about a union between wit and judgment." She was, however, better known as what has been called, in the male sex, "an agreeable rattle"; careless of what she said, and therefore occasionally, more by accident than design, saying a good thing. "Gay, flighty, entertaining and frisky," is the description given of her manner; while so much of her conversation as remains to us is more remarkable for frothy flippancy than for depth. As a matter of course, Johnson frequently encountered her in the society which both affected. Of one such meeting we are told that "Mrs. Cholmondeley, in a high flow of spirits, exhibited some sallies of hyperbolical compliment to Johnson, with whom she was long acquainted, and was very easy." He was quick in catching the *manner* of the moment and answered her somewhat in the style of the hero of a romance, "Madam, you crown me with unfading laurels." Once we learn that he seized her hand and held it close to his eyes wondering at its delicate whiteness, while she exclaimed aside, "I wonder will he give it me again when he has done with it?" But enough has, we think, been said of Mrs. Cholmondeley.*

To readers of 'Vanity Fair' the name of "Mrs. Chapone" must be familiar. They will remember that Miss Pinkerton's academy for young ladies in Chiswick Mall, where Amelia

---

* Those who wish to know more will find a good deal about her in Madame d'Arblay's 'Diary.'

Osborne made the doubtful acquaintance of Becky Sharp, had been "honoured by the presence of *The Great Lexicographer* and the patronage of the admirable Mrs. Chapone"; and indeed the latter eminent person was held by her compeers to fully deserve the epithet by which she is described. Previous to her marriage, Hester Chapone was a Miss Mulso. Her mother had been a remarkably beautiful woman, but did not transmit this characteristic to her daughter. However, what Hester lacked in evanescent charms was made up to her in the more solid endowments of the mind. She began early to give proof of her quality, for at the tender age of nine she produced a romance entitled 'The Loves of Amoret and Melissa.' But her mother, like a sensible woman, promptly put a stop to that kind of thing, and ordained for her a strict course of domestic training highly unfavourable to romance in any shape. Still, Hester's aspirations could not be wholly suppressed, and what little time might be snatched from housekeeping was devoted by her to the acquisition of French, Italian, Latin, Music and Drawing. She also cultivated authorship, and by the time she had attained her twenty-third year Johnson had accepted her as one of the contributors to the *Rambler*. After this she found her way into literary society, and made the friendship of Richardson, with whom she became a pet. In her letters to the author of 'Clarissa,' she signs herself his " ever obliged and affectionate child " ; and he, presumably on account of her sparkling wit, is good enough to call her his " little spitfire." Considering that she was indebted to Johnson for her introduction to the world of letters, she repaid him rather badly by denouncing 'Rasselas,' on its first appearance, as "an ill-contrived, unfinished and uninstructive tale." But retribution was in store for her. Meeting an attorney named Chapone, she fell in love with, and wanted to marry, him. Her father would not at first hear of the match, which had therefore to be postponed. But, true to her principles, whilst waiting for her stern parent's consent, she drew up a 'Matrimonial Creed,' in seven stringent articles of faith, which she inscribed to Richardson and in which the

Dr. Johnson as a Man of Fashion. 113

theory of marriage was, for once, placed upon a sound logical basis. At length her father gave way and the wedding took place; but, strange to say, the Creed did not work well in practice. Having endured its operation with a very ill grace for about a year, Mr. Chapone died, as some averred, of sheer dissent; and his widow never made another proselyte. Henceforth, she devoted herself to the amelioration of a species which she was not fated to increase. In 1772 appeared that work upon which her great reputation was chiefly founded, 'Letters on the Improvement of the Mind'; which impressed the world so much that a great many anxious parents entreated her to undertake the education of their daughters. She took a high place in society at once, and was by special command introduced to Queen Charlotte, who condescended to express a hope that the Princess Royal had profited by the 'Letters'; what the Princess herself hoped, does not appear Many other useful works did this gifted. woman produce—most of which are catalogued in 'The Dictionary of National Bio-

I

graphy'; and her reputation as a Mentor lasted well into the present century, though it is now somewhat forgotten. She is included among "The British Essayists." In person she was not attractive;* but her conversation was better suited to the ordinary intelligence than might have been expected from her works. She had many of the lighter accomplishments, could sketch a portrait, and was held to sing "most exquisitely." Miss Burney describes her as "the most superiorly unaffected creature you can conceive, and full of *agrémens* from good sense, talents, and conversational powers, in defiance of age, infirmities and uncommon ugliness." She spoke warmly of 'Camilla'; but "detected me in some gallicisms, and pointed some out." Despite her galling censure of 'Rasselas,' Johnson and she were very good friends; and she was thought to have great influence with him, as is proved by the applications which were made through her

* *Wraxall* says of her that, "under one of the most repulsive exteriors that any woman ever possessed she concealed very superior attainments and extensive knowledge." But it is only fair to add that her portraits do not bear out his unflattering description.

Mrs. CHAPONE.

by persons who wanted his opinion or advice. Thus, the Earl of Carlisle having written a tragedy, "some of his lordship's friends applied to Mrs. Chapone to prevail on Dr Johnson to read and give his opinion of it, which he accordingly did in a letter to that lady."* We find her name often occurring in Hannah More's correspondence, which is quite enough to show that she was a person in much social request; and she seems to have retained the respect and esteem of all who knew her to the close of her very long life.

If Mrs. Abington was Johnson's theatrical goddess, Kitty Clive was his pet. "Clive, Sir," he used to observe, "is a good thing to sit by, she always understands what you say": and, from all we know of her, the praise was not undeserved. Like her rival, poor Kitty had a very chequered career. About the close of the 17th century there happened to be living in the town of Kilkenny a certain lawyer, named Raftor, who took the unfortunate side when Dutch William was engaged

* BOSWELL.

in driving his father-in-law out of Ireland.
After the battle of the Boyne Raftor escaped
to France, where he remained until, having
been granted a free pardon by Queen Anne,
he took up his residence in London. There
he married a citizen's daughter, and Kitty was
one of the results. Born in penury, the child
was left almost totally uneducated, and to her
dying day never mastered the intricacies of
spelling. But genius will overcome all difficulties; she had a smart tongue and a beautiful voice, by which she managed to attract
the notice of some leading actors who obtained
for her an introduction to Colley Cibber, then
lessee of the Drury Lane Theatre. Cibber
took her in hand; and after she had under-
gone the usual probation, her opportunity
came. Nat Lee's tragedy of 'Mithridates' was
put on the stage, and she was given the part
of a page—"with a song." This was Kitty's
chance; she warbled the song in a manner
which enraptured her audience, and from
that moment her success was assured. The
management soon discovered her particular
bent. Comic parts were then chiefly assigned

to her; and her humour, combined with her ability as a ballad-singer, carried all before it. When she was about twenty the girl met with a stroke of bad luck. She married one Clive, a barrister; and it was not very long before the couple had to separate, though by no fault of hers—for, in all the private relations of life, Kitty was irreproachable. In his preface to 'The Intriguing Chambermaid,' Fielding describes her as "The best wife, the best daughter, the best sister and the best friend" he knew; and nobody ever contradicted him. As an actress, genteel comedy was not her *forte*; but she was acknowledged to be "particularly happy in low humour." Johnson declared: "In the sprightliness of humour I have never seen her equalled. What Clive did best she did better than Garrick; but could not do so many things well: she was a better romp than any I ever saw in nature"; while Goldsmith vowed that, in her line, she excelled any actress he had ever seen in England or abroad. But it was not upon her acting that Kitty most prided herself. She was an accomplished and highly trained musician; and her leaning was not, as might have been expected, to music of the lighter sort, but to the works of the Great Masters. Handel had a very high opinion of her abilities in classic music, as is proved by his entrusting to her the part of Dalilah in his oratorio of 'Samson' when it was first brought out. This fact combined with her remarkable powers of conversation to secure her a good place in society; where, if she did not quite attain the position of Mrs. Abington, she held her own very well, and was always a welcome guest at Strawberry Hill when Horace Walpole entertained the wit and fashion of the day. Though when Kitty was in a good humour nobody could be more charming, she had her defects of temper; indeed, it is sorrowfully owned that she was "too often passionate, cross and vulgar." But some excuse has to be made for her. When Garrick took over Drury Lane Theatre he proved a more exacting manager than Cibber; and certain complications ensued between him and Mrs. Clive in which the fault was not altogether on one side. Kitty, no doubt, tried him very

Mrs. CLIVE.

much; but on the other hand he was sometimes harsh and, to say the least of it, arbitrary. However, to the credit of both parties, they eventually patched up their differences and remained fast friends for life. No one was more pleased than she when Johnson popped his wig into the green-room at "The Lane" and took a chair near hers. It was she who made the remark, "I love to sit by Dr. Johnson, he always entertains me"; and wherever they met, the two managed to gravitate towards each other. Boswell assures us that "he (Johnson) used to converse more with her than with any of the players." In this he admittedly showed his good taste; Kitty being by general consent one of the pleasantest ladies to be found anywhere. Even Davies, who was not over well disposed to her, admits, in his 'Life of Garrick,' that "her company was always courted by women (mark you, *women*) of high rank and education, to whom she rendered herself very agreeable"; and that, after she left the stage, "she is still visited by very distinguished persons of both sexes. Her conversation is a mixture of uncommon vivacity, droll mirth, and honest bluntness." Truly, a delightful combination which worthily fitted her for a place in the Doctor's famous "post-chaise."

It is impossible to omit "Swift's Mrs. Delany" from our collection, though she is somewhat out of place in the group. The lady of whom Burke said, "She was a pattern of a perfect fine lady—a real fine lady—of other days; her manners were faultless, her deportment was all elegance, her speech was all sweetness, and her air and address all dignity," is somehow too fine for the heterogeneous Blue-stocking sisterhood; nor do we find that she identified herself much with their proceedings. Johnson and she, of course, met occasionally; but there does not appear to have been any very close communion between them, and indeed they were rather unsuited to each other. Yet what a delightful companion she must have been with her wonderful flow of old-world talk! As Hannah More says, "She was a Granville, and niece to the celebrated poet Lord Lansdowne. She was the friend and intimate of Swift. She tells a thousand pleasant anecdotes relative to the publication of the *Tatler*. As to the *Spectator*, it is almost too modern for her to speak of it. She was in the next room, and heard the cries of alarm when Guiscard stabbed Lord Oxford. In short, she is a living library of knowledge; and time, which has so highly matured her judgment, has taken very little from her graces or her liveliness." When this was written she was over eighty and almost blind, yet happy were they who could contrive to secure an interview with the charming old lady, who was as good as she was distinguished, who retained "all that tenderness of heart which people are supposed to lose, and generally do lose, at a very advanced age"; who had "no dread of death but what arose from the thought how terribly her loss would be felt by one or two dear friends." We all know how fond King George and his Queen were of her; how the former honoured her with a royal osculation, and how the two eventually found that they could not part with her. Miss Burney narrates that she and Mrs. Chapone one day called upon her at her cosy house in St.

James's Place ; and this is how she draws her for us :—

"Mrs. Delany was alone in her drawing-room, which is entirely hung round with pictures of her own painting, and ornaments of her own designing. She came to the door to receive us. She is still tall, though some of her height may be lost; not much, however, for she is remarkably upright. She has no remains of beauty in feature, but in countenance I never but once saw more, and that was in my sweet maternal grandmother. Benevolence, softness, piety, and gentleness are all resident in her face. . . . Mrs. Chapone presented me to her, and taking my hand, she said,—'You must pardon me if I give you an old-fashioned reception, for I know nothing new,' and she saluted me."

A single strong touch would be out of keeping with the quiet harmony of this picture, and Johnson's brusque method is therefore inapplicable. We have no record of conversations between him and Mrs. Delany ; nor do we think there were many.

There was a Mrs. Fitzherbert—not to be confounded, if you please, with *the* Mrs. Fitzherbert who got mixed up with a certain prince at a southern watering-place ; but a

Mrs. DELANY.

lady of unsullied reputation. Johnson had a liking for her, as the daughter of his old friend Meynell who had such a happy knack of saying unpremeditated good things; but he also admired her greatly for herself, and is even reported to have said that she had "the best understanding he ever met with in any human being." Unfortunately, we have no means of judging how far this opinion was justified, except a solitary observation which is said to have fallen from her in reference to Homer's description of the shield of Achilles, "He may hold up that SHIELD against all his enemies." Johnson used to repeat this as a very fine saying; but we think we have heard finer. Mrs. Fitzherbert was rather fortunate in her husband, of whom the Doctor said that he never knew a man who was so generally acceptable. "He made everybody quite easy, overpowered nobody by the superiority of his talents, made no man think worse of himself by being his rival, seemed always to listen, did not oblige you to hear much from him, and did not oppose what you said"—young gentlemen who wish to get on in the world will please note. Still, there were drawbacks. Mrs. Thrale tells us that Johnson "spent much of his time with Mrs. Fitzherbert, of whom he always spoke with esteem and tenderness, and with a veneration very difficult to deserve. 'That woman,' said he, 'loved her husband as we hope and desire to be loved by our guardian angel ... her first care was to preserve her husband's soul from corruption; her second, to keep his estate entire for their children; and I owed my good reception in the family to the idea she had entertained that I was fit company for Fitzherbert, whom I loved extremely.' 'They dare not,' said she, 'swear and take other conversation liberties before *you*.' ... She stood at the door of her paradise in Derbyshire, like the angel with a flaming sword, to keep the devil at a distance. But she was but mortal, poor dear—she died, and *her husband felt at once afflicted and released.*" On Mrs. Thrale asking if she was handsome, the reply was that "She would have been handsome for a queen. Her beauty had more in it of majesty than of attraction; more of the dignity of virtue than the vivacity of wit." A stately, somewhat awful picture—pure, cold, serene, and slightly out of place with its surroundings.

In the year 1749 Mr. David Garrick married the young and charming Mdlle. Violetti—"a young lady," says his biographer, Davies, "who to great elegance of form, and many polite accomplishments, joined the more amiable virtues of the mind." Considering the relation in which Johnson and Garrick stood to each other, it was inevitable that the former should be to a certain extent her friend ; but it is doubtful whether there was ever any very real communion between the two, though their acquaintance was kept up while Johnson lived. Perhaps, the circumstance of Mrs. Garrick being a foreigner stood between her and the Doctor, whose estimate of foreigners, as a class, was not high ;* or, perhaps, the lady was too gentle and refined

* " For anything I see, foreigners are fools," was a saying of Meynell's adopted by Johnson. Reynolds declared that "he (J.) considered every foreigner a fool till they had convinced him to the contrary."

for those bracing controversies which he loved; but at any rate there is no record of any interchange of sentiment between them, such as we find in the case of his other friends —not a single remark by, or to, the lady which even Boswell cared to preserve. But there were many reasons why their intimacy should be kept up, even when Garrick was no more. His widow was one of the leaders of society. She had a large income, and a spacious house in the Adelphi, where she entertained the best people in a handsome way. For, from the time of her first arrival in England, she had been warmly taken up by people of rank, although she was then only a danseuse at the Haymarket Theatre. The Earl and Countess of Burlington had her to stay with them, did their best to advance her interests in every way, and were actually believed to have settled £6000 on her at her marriage. Everybody liked her; and Miss Burney, who made her acquaintance soon after Garrick's death, explains why. " She was extremely kind and obliging. She looks very well and very elegant. She was cheerfully

MRS. GARRICK.

grave, did not speak much, but was followed and addressed by everybody." She always spoke with a decidedly foreign accent, which Miss Burney mimics, and was given to embracing people—both, things attractive in their way. Boswell tells us how, " On Friday, April 20 (1781), I spent with him (Johnson) one of the happiest days that I remember to have enjoyed in the whole course of my life. Mrs. Garrick, whose grief for the loss of her husband was, I believe, as sincere as wounded affection and admiration could produce, had this day, for the first time since his death, a select party of his friends to dine with her. The company was Miss Hannah More, who lived with her, and whom she called her chaplain; Mrs. Boscawen, Mrs. Elizabeth Carter, Sir Joshua Reynolds, Dr. Burney, Dr. Johnson, and myself. We found ourselves very excellently entertained at her house in the Adelphi, where I have passed many a pleasing hour with him 'who gladdened life.' She looked very well, talked of her husband with complacency, and while she cast her eyes upon his portrait, which hung over the chimney-piece, said that 'death was

now the most agreeable object to her.'"*
This "agreeable object" was not to be realised
very speedily; for she survived Garrick forty-
three years, and in fact lived to be ninety-
eight or ninety-nine, popular, and deservedly
so, to the last. If she never said or did anything very remarkable; if she was the only
one of Johnson's friends to whom, as far as
we know, none of his "good things" were
addressed, she was not the less what her sex
are wont to describe as "a most delightful
person." And so she is pourtrayed to us,
graceful, elegant, charming, kind; a pretty
picture, somewhat weak in drawing and de-
ficient in colour—hardly "strong" enough, in
fact, for the Johnson gallery, but the omission
of which would leave a gap in that collection.

Mrs. Gastrell was the widowed sister of that
"Molly Aston" whom we have met before,
and lived at Stowhill, near Lichfield. While
Johnson and Boswell were staying at the
latter place, Johnson one day "walked away
to dinner there, leaving me by myself without

* The full account of this dinner given by Boswell is
well worth reading, but too lengthy for insertion here.

any apology"; but soon came a note in the Doctor's handwriting: "Mrs. Gastrell, at the lower house on Stowhill, desires Mr. Boswell's company to dinner at two," and the delighted Boswell "had here another proof how amiable his character was in the opinion of those who knew him best." She must have been an agreeable lady, for when Boswell afterwards visited Lichfield on his own account, he made a point of paying his respects to Mrs. Gastrell, "whose conversation I was not willing to quit." It is, however, painful to be reminded that her deceased husband was "the clergyman who, while he lived at Stratford-upon-Avon, where he was the proprietor of Shakespeare's garden, with Gothic barbarity cut down his mulberry-tree and, as Dr. Johnson told me, did it to vex his neighbours. His lady, I have reason to believe, participated in the guilt of what the enthusiasts for our immortal bard deem almost a species of sacrilege." After this revelation, the less said of Mrs. Gastrell the better; but Johnson always preserved his friendship for her, as indeed for all who belonged to "Molly's" family, with whom so many tender memories were associated.

Mrs. Knowles, "the ingenious Quaker lady," stands out prominently among Johnson's friends for the reason that, whenever they met, she gave him battle and rarely sustained defeat. As Miss Mary Morris, she had earned distinction not only by her wit and beauty, but by her proficiency in a species of worsted-work, which Johnson called "sutile pictures" —Mrs. Thrale, whether by accident or design, describes them as "*futile* pictures"—and in which she was commissioned to execute portraits of George the Third and his family. She married a Dr. Knowles, who died in 1784, leaving her remarkably well provided for; but the husband is quite thrown into the shade by his brilliant wife. Celebrated for her conversational powers, she soon came into contact with Johnson on all manner of subjects, but chiefly on questions of religion and the disabilities under which her sex then laboured. The contest was fair enough until she took the mean advantage of publishing, in 1776, 'A Compendium of a Controversy on Water-

Mrs. KNOWLES.

Baptism ; also, a Dialogue between Dr. Johnson and Mrs. Knowles respecting the Conversion to Quakerism of Miss Jane Harvey.' In this "compendium" the Doctor is certainly represented as having come off second best; but, angry though he was at the betrayal, he could not help laughing at the following touch of nature in the final paragraph. His fair adversary is made to admonish him thus: "I hope, Doctor, thou wilt not remain unforgiving, and that you (he and Miss Harvey) will renew your friendship, and joyfully meet at last in those bright regions where pride and prejudice* can never enter." DOCTOR: "Meet her! I never desire to meet fools anywhere." But despite hard knocks on both sides, the two combatants had a profound respect for each other. Mrs. Knowles watching him once engaged with a book, which he seemed to "read ravenously, as if he devoured it," said, "He knows how to read better than anyone; he gets at the substance of a book directly; he tears out the heart of it." And

* It is supposed that Miss Austen took the title of her novel, 'Pride and Prejudice,' from this passage.

when they were battling over the question, why Quakers should take it upon them to style everybody *friends*, Mrs. Knowles having made a dexterous application of a certain text to support her particular view, Johnson replied (with eyes sparkling benignantly), "Very well, indeed, Madam ; you have said very well!" This was true chivalry—two champions, skilful of fence, actuated by no petty rivalry, but each keen to appreciate the science of the other, and to cry "Well done!" at a good stroke.

Mrs. Lenox was born in 1729 at New York, of which her father was then Lieutenant-Governor, but came to England, as a child, with prospects which were not fulfilled—for at his death she was left wholly unprovided for. Having tried to be an actress and failed, she took to literature, in which she was rather more successful. In 1748 she changed her maiden name of Ramsay for that of Lenox ; but her marriage proved unfortunate, and she was compelled to rely upon her pen as a means of support. In this way she made the acquaintance of Johnson, whose admiration of

Dr. Johnson as a Man of Fashion. 133

the lady's appearance and virtues, combined with pity for her misfortunes, disposed him to rate her talents somewhat extravagantly. Indeed, he cited her as an instance of *talent* in his 'Dictionary'; and, to celebrate the publication of her novel, 'Harriot Stuart,' invited her to supper at his club. One of the dishes on this festive occasion was, we are informed, "an enormous apple-pie, which he stuck with bay-leaves; and he had prepared for her a crown of laurel with which he encircled her brows." Such adulation from the Monarch of Letters turned the poor lady's head, who began to give herself airs, with the result that, as we learn from Mrs. Thrale, "while her books are [generally approved, nobody likes her." A plot was formed to damn her comedy 'The Sister' on the first night of its performance, and succeeded, though the piece is stated to have been well written. One of her novels, 'The Female Quixote,' is decidedly clever, and elicited the praise of so good a judge as Fielding; but she rarely conquered outside the realms of fiction; and her 'Shakespeare Illustrated,' in which she collected the various sources from which Shakespeare was supposed to derive his plots, adding comments of her own, proved a dismal failure. Johnson, as we have seen before, at one time ranked her as superior to all the clever women of his acquaintance,* but found nobody to agree with him. However, he always assisted her in every way he could ; and Boswell tells how, " the first effort of his pen in 1775 was 'Proposals for Publishing the Works of Mrs. Charlotte Lenox in three volumes quarto.' In his diary, January 2, I find this entry : ' Wrote Charlotte's proposals.' " After her kind friend's death the poor lady fell into great penury, and became a pensioner on the Royal Literary Fund. She deserved a better fate, for she had worked very hard to achieve success, and had considerable talent in her way.

When it is said of Lady Lucan that she was wife of the first Earl of Lucan, what more is required to fix her ladyship's position in society? Boswell alludes with becoming reverence to the intimate terms upon which Johnson stood with her and her noble spouse.

* See note to p. 108.

Mrs. LENOX.

Dr. Johnson as a Man of Fashion. 135

"No author by profession," he assures us, "ever rose in this country into that personal notice which he (Johnson) did. In the course of this work a numerous variety of names have been mentioned, to which many might be added. I cannot omit Lord and Lady Lucan, at whose house he often enjoyed all that an elegant table and the best company can contribute to happiness; he found hospitality united with extraordinary accomplishments, and embellished with charms of which no man could be insensible." The "extraordinary accomplishments" are reduced by Horace Walpole to the single one, that her ladyship had "an astonishing genius for copying whatever she sees. The pictures I lent her from my collection, and some advice I gave her, certainly brought her talents to wonderful perfection in five months; for, before, she painted in crayons as ill as any lady in England." He, however, takes occasion to add, "she models in wax, and has something of a turn towards poetry"; but laments that "her prodigious vivacity makes her too volatile in everything, and my lord follows wherever she leads." This vivacious lady delighted much in Johnson's company, when at all *distraite*. According to Rogers, the poet, "she would say to her daughter, afterwards Lady Spencer, 'Nobody dines with us to-day; therefore, child, we'll go and get Dr. Johnson.' So they would drive to Bolt Court, and bring the Doctor home with them." Of course, he was a prominent figure at her ladyship's assemblies. Walpole writes: "I saw Dr. Johnson last night at Lady Lucan's, who had assembled a Blue-stocking meeting in imitation of Mrs. Vesey's *Babels*.* It was so blue, it was quite Mazarine blue. Mrs. Montagu kept aloof from Johnson like the West from the East." And, again, he writes that, "at a Blue-stocking meeting held by Lady Lucan, Mrs. Montagu and Dr. Johnson kept at different ends of the chamber, and set up altar against altar there." †

* It was Walpole himself who originated this nickname for Mrs. Vesey's *Blue* parties. "She collects all the graduates and candidates for fame, when they vie with one another until they are as unintelligible as the good folks at Babel."

† For an account of this feud, see p. 161.

Dr. Johnson as a Man of Fashion. 137

If Mrs. Catherine Macaulay was less versatile than Mrs. Montagu, less learned than Mrs. Carter, and less brilliant than Mrs. Thrale, she was superior to them all, perhaps indeed to any woman of her time, as an author and thinker. Mr. Lecky pronounces her "the ablest writer of the new Radical school"; and from the estimation in which she was held by its chief exponents, it is probable that Mr. Lecky does not say too much. Her tendency to Radicalism existed from her earliest years; for being by her father's wish educated privately, she had made a special study of Roman history, and this imbued her with an enthusiasm for liberty which coloured all her future life. By the usual channels she found her way into the ranks of literature; but did not take any prominent position therein until after her marriage to Dr. Macaulay, when the first volume of her 'History of England from the Accession of the Stuarts' appeared, which created a great sensation, making her many friends and more enemies. The virulence of her critics was astounding. Not content with denouncing her views, they attacked her personal appearance—though without sufficient reason, as she is said to have been "tall and of a good figure." At this time she was a little over thirty, having been born in 1731, and ought, therefore, to have been looking her best; but the Tory press was then accustomed to spare neither man nor woman in its wrath, and handled her very scurvily. When she had been married about six years Dr. Macaulay died; and she was left free to complete her 'History,' which secured her the admiration of all unprejudiced critics. Horace Walpole joined Gray the poet in pronouncing it "the most sensible, unaffected, and best history of England that we have had yet," and Mirabeau was anxious that it should be translated into French. Meantime, her literary pursuits were not allowed to interfere with her pleasures. Being fond of gaiety, she went to live at Bath in 1774, when the practice of good living had, perhaps, induced those symptoms which "the waters" are supposed to alleviate; and there she made the acquaintance of Dr. Wilson, the rector of St. Stephen's, Walbrook, who had a house in Alfred Street, Bath, which, together with its furniture and a valuable library, he placed at her full disposal, and in fact never deprived her of, though he had afterwards cause to repent the gift. In 1775 she visited Paris, and was received there with so much honour that she revisited it in 1777, when all the celebrities of that capital flocked round the distinguished priestess of Liberty, and Madame Roland avowed her own aspirations to become "*la Macaulay de son pays.*" But French frivolity is contagious; and on her return to England Mrs. Macaulay's admirers hardly recognised their old friend in the painted and bedizened harridan who leered upon them from behind her rouge. The genial Mr. John Wilkes describes her as "painted up to the eyes, and looking as rotten as an old Catherine pear"—observe the elegance of the pun. Dr. Johnson, who naturally was not well disposed to a "vile Whig" historian, opined that she was "better employed in reddening her own cheeks than in blackening other people's characters." In fact, everybody was more or less angry with her; and her subsequent proceedings did not abate their resentment. For in 1778, that is when forty-seven years old (she was born in 1731), she amazed all Bath by marrying a young fellow of twenty-one, a Mr. William Graham, brother to a well-known quack doctor of the day. Her benefactor, the Rev. Dr. Wilson, was furious. He could not take from her the house in Alfred Street, but he gave orders that a marble statue which he had erected to her in the chancel of St. Stephen's, Walbrook, should be removed, and that a vault which he had kindly provided for her use in the sacred edifice should be sold. After her marriage she left Bath and, in 1784, together with her husband, whom she had by this time converted into a clergyman, visited North America, where she was for some time the guest of Washington at Mount Vernon. Returning to England she took up her residence at Binfield, where a monument erected to her in the parish church informs us that she died in 1791. Beyond all question, she was possessed of great talents and indomitable energy. Works upon all manner

Dr. Johnson as a Man of Fashion. 141

of subjects issued from her pen with wonderful rapidity, and though she never repeated the success achieved by her 'History of England,' her reply to Burke's 'Thoughts on the Cause of the Present Discontent' and her 'Modest Plea for the Property of Copyright' deserve more praise than they got. Her Radical views militated against her success as a woman of fashion, and exposed her to many a rebuff. The story of how Johnson put her down at her own table, though well-known, will bear quoting again:—

" Sir, there is one Mrs. Macaulay in this town, a great Republican. One day when I was at her house I put on a very grave countenance, and said to her, ' Madam, I am now become a convert to your way of thinking. I am convinced that all mankind are on an equal footing; and to give you an unquestionable proof, Madam, that I am in earnest, here is a very sensible, civil, well-balanced fellow-citizen, your footman : I desire that he may be allowed to. sit down and dine with us.' I thus, Sir, showed her the absurdity of the levelling doctrine. She has never liked me since. Sir, your levellers wish to level *down* as far as themselves; but they cannot bear levelling *up* to themselves. They would all have some people under them; why not then have some people above them?"

It is likely that Mrs. Macaulay and the Doctor seldom met without some such dispute, and that the acquaintances of both took pleasure in their encounters. A suspicion of this may have crossed the Doctor's mind and nettled him; for we have the following entry in Boswell's 'Journal':—

"On Monday, September 22, when at breakfast, I unguardedly said to Dr. Johnson, 'I wish I saw you and Mrs. Macaulay together.' He grew very angry; and, after a pause, while a cloud gathered on his brow, he burst out, 'No, Sir, you would not see us quarrel, to make you sport. Don't you know that it is very uncivil to *pit* two people against one another?' Then checking himself, and wishing to be more gentle, he added, 'I do not say that you should be hanged or drowned for this; but it *is* very uncivil.'"

Though secretly admiring her talents, Johnson persisted in quizzing her about her socialist proclivities; and the lady resented this so much, that waggish friends suggested the two had better get married in order to have the fullest opportunity for quarrelling

Mrs. MACAULAY.

perpetually. Johnson rather enjoyed the joke, as appears from the following conversation between him and Boswell :—

"BOSWELL: 'Sir, you'll never make out this match, of which we have talked, with a certain political lady, since you are so severe against her principles?' JOHNSON: 'Nay, Sir, I have the better chance for that. She is like the Amazons of old; she must be courted by the sword. But I have not been severe upon her.' BOSWELL: 'Yes, Sir, you have made her ridiculous.' JOHNSON: 'That was already done, Sir. To endeavour to make *her* ridiculous, is like blacking the chimney.'"

Poor Mrs. Macaulay, like others of the Blue-stocking sisterhood, outlived her reputation, and her later works did not serve to increase it. She lived to be eighty; and though she was known to the last as "the celebrated female historian," people grew to be rather chary of accepting those cards of invitation which were signed, " Catherine Macaulay, *At Home to the Literati.*" Nor did they lament her over much when she died.

In the year 1765 there happened to be a young English lady, a Miss Riggs, with a great deal of money to her fortune, and by an odd coincidence there also happened to be an Irish military gentleman (on half-pay), named Miller, who wanted money sorely. As luck would have it, the two met, and after the manner of Irishmen Mr. Miller carried off the heiress. He proved a good-natured husband, and spent most of her fortune in building for his wife a sumptuous villa at Batheaston near Bath, where they enjoyed themselves so much that the money gave out in a wonderfully short time, and they were fain to shut up their villa and go abroad to retrench. While they were making the grand tour, Mrs. Miller, who had a pretty gift that way, wrote home three volumes full of 'Letters from Italy,' which, being duly published, hit the taste of the public, though Horace Walpole avers that the fair authoress "did not speak one word of French or Italian right all through them." Having got over their money difficulties somehow, the pair returned to Batheaston; and Mr. Miller being converted into an Irish baronet, his wife henceforth became known as Lady Miller, and

Dr. Johnson as a Man of Fashion. 145

resolved to be still more famous as a Bluestocking. With this object in view, she instituted a literary *salon*, in imitation of the Della Cruscans, and invited all persons of wit and fashion in Bath to assemble once a fortnight at her house. While in Italy the Millers had secured an antique vase said to have been found at Cicero's villa, and hence called the *Tully* vase. This was now placed upon a modern altar at Batheaston, and when the fortnightly entertainments came round was decorated with flowers and ribbons and garlands of laurel. Each guest was then invited to deposit in the vase a composition in verse; a committee was appointed to determine what were the three best of such deposits; and this being ascertained, Lady Miller, got up as a pagan priestess, decorated the authors thereof with wreaths of myrtle. The priestess is described by Miss Burney about this time as "a round, plump, coarse-looking dame of about forty"—not a very poetic person to the eye, perhaps; but then, with *such* a soul! When the noise of these doings reached London there was a considerable flutter amongst the Blue-stockings, who vowed that Lady Miller was bringing discredit upon the order, and were inclined to hold her ladyship up to ridicule ; but after she came up to town, what with her wealth and her prestige and her exclusiveness, she soon became a person of mark. "Do you know now," writes Miss Burney in 1780, "that notwithstanding Batheaston is so much laughed at in London, nothing here is more tonish than to visit Lady Miller, who is extremely curious in her company, admitting few people who are not of rank or fame, and excluding all those who are not people of character very unblemished." When, only a year afterwards, her ladyship died at the age of forty-one—being almost the only Blue-stocking who died so young—people said it was a "judgment" on her.

So recently as 1840 there died in New Burlington Street, London, an old lady who had been the intimate friend of Dr. Johnson, and was "The last of the Blue-stockings"; who, according to the late Mr. Hayward, was "the first Englishwoman of rank who threw

Dr. Johnson as a Man of Fashion. 147

open her house to literature, or made intellectual distinction a recognised passport to society." This lady had burst upon the world nearly a century before as the youngest child and only surviving daughter of John Monckton, first Viscount Galway, whose house in Charles Street, Berkeley Square, she made the rendezvous of clever people. For the Honourable Miss Monckton adored talent from her earliest years, dabbled in literature herself; and in conjunction with her mother, Lady Galway, held what was known as "the finest bit of Blue" in all London. At their house was to be seen everybody who was worth seeing, and she remained an ardent lion-hunter to the end of her days. Of course, Johnson was a welcome guest in Charles Street, and through him she became acquainted with Mrs. Thrale and Miss Burney, the latter of whom sketches her thus:—

" Miss Monckton is between thirty and forty, very short, very fat, but handsome; splendidly and fantastically dressed, rouged not unbecomingly, yet evidently and palpably desirous of gaining notice and admiration. She has an easy levity in her air, manner, voice and discourse that speaks all within tobe comfortable; and her rage of seeing anything curious may be satisfied, if she pleases, by looking in a mirror."

Then follows a description of an assembly at Lady Galway's:—

"There was not much company, for we were very early. Lady Galway sat at the side of the fire, and received nobody. . . . Such part of the company as already knew her made their compliments to her where she sat, and the rest were never taken up to her, but belonged wholly to Miss Monckton. Miss Monckton's own manner of receiving her guests was scarce more laborious; for she kept her seat when they entered, and only turned round her head to nod it, and say, 'How do-do?', after which they found what accommodation they could for themselves. . . . Dr. Johnson was standing near the fire, environed with listeners. .... . The company in general were dressed with more brilliancy than at any rout I ever was at, as most of them were going to the Duchess of Cumberland's."

Miss Monckton would not tolerate card-playing at her parties; but people with Johnson, Burke and Reynolds to talk to had not much to complain of. In other respects she made them pretty comfortable, and, as her observant guest notes, was "far better at her own house than elsewhere." How Johnson and she comported themselves together is painted for us by Boswell :—

"Her vivacity enchanted the Sage, and they used to talk together with all imaginable ease. A singular instance happened one evening when she insisted that some of Sterne's writings were very pathetick. Johnson bluntly denied it. 'I am sure' (said she) 'they have affected *me.*' 'Why' (said Johnson, smiling and rolling himself about) 'that is because, dearest, you are a dunce.' When she some time afterwards mentioned this to him, he said with equal truth and politeness, 'Madam, if I had thought so, I certainly should not have said it.'"

Miss Burney gives another instance :—

"'But indeed, Dr. Johnson,' said Miss Monckton, 'you *must* see Mrs. Siddons. Won't you see her in some fine part?'

"'Why, if I must, Madam, I have no choice.'

" 'She says, Sir, she shall be very much afraid of you.'

"'Madam, that cannot be true.'

"'Not true,' cried Miss Monckton, staring; 'yes it is.'

"'It *cannot* be, Madam.'

"'But she said so to me: I heard her say it myself.'

"'Madam, it is not *possible*. Remember therefore, in future, that even fiction should be affected by probability.'

"Miss Monckton looked all amazement, but insisted upon the truth of what she had said.

"'I do not believe, Madam,' said he warmly, 'she knows my name.'"

But despite their sparring-matches, Johnson and she remained fast friends; and when the old man became too infirm to go into society she used to visit him at his own house, whither Hannah More mentions going in her company. When forty years of age, it occurred to her that she had remained single long enough; so she became Countess of Cork, with great enhancement to her splendour of appearance. Jekyll describes her at this period as being "like a shuttlecock, all cork and feathers." The older she grew the stronger became her passion for entertaining notabilities. Lord Beaconsfield, who knew her well, is said to have described her accurately as "Lady Bellair" in 'Henrietta Temple'; and Dickens is thought to have had her in his eye when drawing "Mrs. Leo Hunter" in 'Pickwick.' She lived to be ninety-four.

The biography of Mrs. Hannah More is too serious a matter to be handled at any length in these pages.* We first hear of her as a remarkably precocious child who at the age of four, or thereabouts, had learned to read, simply from being allowed to be present while her elder sisters were pursuing their education. By the time she was eight she found a craving for classical history, and her father tried her with small doses of Latin and mathematics, but grew "frightened at his own success." Next, she picked up a knowledge of French by talking to some officers of that nation who happened to be at Deal on parole. Then, her sisters having opened a school, she waylaid their Italian, Spanish and Latin masters, thereby making herself proficient in these languages—altogether a very

* For a full and true account of this lady the reader is referred to Mr. Leslie Stephen's exhaustive monograph in the 'Dictionary of National Biography.'

pushing young lady. Meanwhile, she had been attempting essays and other small literary ventures; and was not long in publishing an English adaptation of Metastasio's 'Regulus,' which, under the title of 'The Inflexible Captive,' was brought out on the provincial stage. At twenty-two she received an offer of marriage, which she accepted; but the gentleman could not make up his mind for the final plunge, and the engagement was broken off—the consequence being that she determined to 'forswear matrimony in future. In 1777 she visited London; saw Garrick in 'Lear,' and wrote to one of her friends a fervent description of the effect which his acting had produced upon her. The friend judiciously showed this letter to Garrick, whereupon he and his wife sought out the writer, and had her to stay with them for some months at their house in the Adelphi. While there, she made the acquaintance of Burke and Reynolds, by the latter of whom she was introduced to Dr. Johnson; and from this point we must confine ourselves to the friendship which resulted. She soon became

HANNAH MORE.

Dr. Johnson as a Man of Fashion. 153

a favourite with the impressionable sage, as was evinced by his lavishing upon her such terms of endearment as "Little fool," "Love," and "Dearest"; but although flattery was meat that his soul delighted in, she thrust it upon him in such profusion that his gorge rose thereat, and he was obliged to request her, both personally and by deputy, to stay her hand. Miss Burney gives us to understand that this "little rift within the lute" disclosed itself at once :—

"When she was introduced to him, not long ago, she began singing his praise in the warmest manner, and talking of the pleasure and the instruction she derived from his writings, with the highest encomiums. For some time he heard her with that quietness which a long course of praise has given him; she then redoubled her strokes, and, as Mr. Seward calls it peppered still more highly; till, at length, he turned suddenly to her, with a stern and angry countenance, and said, 'Madam, before you flatter a man so grossly to his face, you should consider whether or not your flattery is worth his having.'"

This rebuff might well have discouraged a neophyte; but it required more to repel the ardent Hannah; and we find that Johnson was actually compelled to ask Miss Reynolds to remonstrate with his too zealous worshipper, *teste* Boswell :—

" Talking of Miss ——, a literary lady, he said, 'I was obliged to speak to Miss Reynolds, to let her know that I desired she would not flatter me so much.' Somebody now observed, 'She flatters Garrick.' JOHNSON : ' She is in the right to flatter Garrick. She is in the right for two reasons ; first, because she has the world with her, who have been praising Garrick these thirty years; and secondly, because she is rewarded for it by Garrick. Why should she flatter *me*? I can do nothing for her. Let her carry her praise to a better market.'"

At last, finding perhaps that such methods were unavailing, the persecuted Doctor took to fighting her with her own weapons ; and it must be admitted that he showed himself, at least, a match for her in this kind of warfare. When the poem of 'Bas Bleu' appeared, he declared loudly that she was "the most powerful versificatrix in the English language," that there was "no name in poetry that might not be glad to own her." And when at a dinner-table where both were present one of the guests happened to turn the conversation to poetry—" Hush, hush!" said Johnson, looking towards the fair Hannah, "it is dangerous to say a word of poetry before *her;* it is talking of the art of war before Hannibal." Truly, the man was invincible.

But, having learned to respect each other's prowess, they became fast friends, and the Doctor had a genuine respect for her great worth, undoubted talents, and wonderful energy. He bracketed her with Mrs. Carter and Miss Burney in the eulogy, "three such women are not to be found"; and to the day of his death she was kind and affectionate to the old man, paying him many a long visit at his house when he grew too infirm to leave it. As her fame became noised abroad by the publication of her dramatic and other writings she rose to a more and more prominent place among the "Blue-stockingers," at whose assemblies the Doctor and she encountered, to the huge delight of the bystanders; and at private houses, like that of Mrs. Garrick, where she may be said to have been a permanent resident, they had many a friendly bout. When

Garrick and he were both dead, her bent became deeply religious ; and she may be said to have forsworn the gaieties of life, though still preserving a few worldly friendships, such as that of Horace Walpole. Gradually, she became more and more identified with the cause of religion and philanthropy, Newton and Wilberforce figuring as her chief friends ; her publications were now altogether of a religious character, and circulated in millions. After a busy and most useful life she died, in 1833, aged eighty-eight, and leaving behind her the sum of £30,000, the harvest of her pen.

At the north-west corner of Portman Square stands a large detached house, surrounded by much ground and having a great Ionic *porte cochère* in front,* which was once the residence of her whom Dr. Doran has selected as a typical "Lady of the Last Century." Among the literary women of her own day Mrs. Montagu held a position corresponding to

* Now No. 22, Portman Square ; but really situated in Upper Berkeley Street.

Mrs. MONTAGU.

that of Dr. Johnson among the men; in a word, she was Queen of the Blue-stockings, and entitled to that eminence by her great wealth and lavish display of it no less than by her reputation for wisdom. She was, indeed, a very magnificent person. Rumour assigned to her, at the very lowest estimate, a fortune of ten thousand a year, which she spent royally. Her presence was imposing, and she availed herself of the resources of art to enhance its attractions. Her manner was gracious, as befitted a sovereign; and if she took care always to lead the conversation where she was present, no one disputed that she was worth listening to. Nor did she scorn the trifling arts and accomplishments practised by weaker members of her sex. Wraxall in effect tells us : " Mrs. Montagu was the *Madame du Deffand* of London ; and her influence rested on more solid foundations than those of mere intellect, for she was very wealthy and extremely hospitable. About this time (1776) she was over sixty, though she looked much younger and paid considerable attention to her dress which, though not particularly tasteful, was very costly, being loaded with diamonds and ribbons even when she was four-score. Her voice was unmusical, and her manner dictatorial. She talked well, and loved to talk—always leading the conversation." * Mrs. Delany gives a still more unexpected description of the august lady— " She was handsome, fat and merry ; and, being very fond of dancing, acquired the *sobriquet* of ' Fidget.' " Before the palace in Portman Square was built she had lived in Hill Street, Mayfair, her rooms in which are thus depicted by the last-named writer—the date being 1773, when Mrs. Montagu was fifty-seven. " If I had paper and time I could entertain you with the account of Mrs. M.'s *Room of Cupidons*, which was opened with an assembly for all the foreigners, the literati, and the macaronis of the present age. Many and sly are the observations. How such a genius, at her age, and so circumstanced (Mr. M. had recently taken his upward flight), could think of painting the walls of her dressing-room with bowers of roses

* Wraxall's description, of which the above is a summary, will be found at pp. 137-9 of the ' Historical Memoirs,' Vol. I.

and jessamines entirely inhabited by little cupids in all their little wanton ways, is astonishing." She was indeed a gorgeous potentate, and a kindly one too; for not only did she hold public breakfasts and Bluestocking parties under her hospitable roof in Portman Square, but every May-day she used to entertain there the poor chimneysweeps of London. For nearly fifty years she maintained a practically undisputed supremacy, as hostess, in the intellectual society of the capital; and to her assemblies was pre-eminently due the epithet of "Bluestocking," if those of Mrs. Vesey and Mrs. Ord were similarly designated. At these reunions card-playing was not permitted, the guests being expected to amuse themselves with discussions upon literary topics; but her suppers made amends for all. Hannah More eulogised her in 'Bas Bleu' for having rescued fashionable life from the tyranny of whist and quadrille. Pulteney, Earl of Bath, declared that a more perfect human being was never created; and Burke, upon being told of this, exclaimed, "I do not think he said a word too much." At one period (alas, too brief!) Johnson himself used thoroughly to enjoy conversing with her. "She diffuses more knowledge than any woman I know, or indeed almost any man," he once gallantly observed to Mrs. Thrale. And when she visited Paris after the peace of 1763, we are told that "she perfectly astounded the Parisians by her pecuniary and mental resources." But the climax of her magnificence was reached when, in 1790, she entertained the King and Queen, and as many as seven hundred guests, at a breakfast in Portman Square. She ought to have died then; but she lived ten years onger, and it is sad to read that "her old age was not venerable, owing to her weakness for finery." Samuel Rogers alludes to her at this period as "a compendium of art"; yet in spite of her weakness, those who knew her well loved her well. As a writer, she is best known by her 'Essay on Shakespeare'; but it must be confessed that in this respect she is disappointing, though her letters will repay perusal.

How a breach came between her and Johnson is only partially explained to us. The chances are that he felt jealous of her supremacy, being one of those arbitrary monarchs who could tolerate no rival near the throne. At any rate, events were precipitated by the glory which accrued to Mrs. Montagu on the publication of her 'Essay.' He loudly proclaimed that it did not contain one atom of true criticism ; and in this opinion, which we fear is only too correct, he was supported by Beauclerk and Mrs. Thrale. "One day, at Sir Joshua's table, when it was related that Mrs. Montagu in an excess of compliment to the author of a modern tragedy had exclaimed, ' I tremble for Shakspeare,' Johnson said, 'When Shakspeare has got —— for his rival, and Mrs. Montagu for his defender, he is in a poor way indeed.'"* No doubt, some kind friend made Mrs. Montagu aware of the Doctor's bad taste ; but no actual mischief resulted, until a censorious remark in his ' Life of Lyttelton ' † caused her much annoyance. Still she made no sign ; but her rival

* Boswell.
† She had contributed some dialogues to Lyttelton's 'Dialogues of the Dead.'

was eager for the fray. On being informed that "she had bound up Mr. Gibbon's 'History' without the last two offensive chapters; for that she thought the book so far good, as it gave, in an elegant manner, the substance of the bad writers *medii ævi*, which the late Lord Lyttelton advised her to read," Johnson broke in with, " Sir, she has not read them; she shows none of this impetuosity to me; she does not know Greek, and, I fancy, knows little Latin. She is willing you should think she knows them; but she does not say she does." Things went on from bad to worse. Johnson growled and stormed and behaved himself very badly. But what did Mrs. Montagu do?—exactly what a wise woman of the world should have done; she began to ignore him, asking him to dinner just as usual, indeed, but taking very little notice of him when he came. Johnson felt the slight as few other men would. He; *he*, the celebrated Dr. Johnson, to be treated as a mere nobody!—it was intolerable. He affected indeed to make light of it, but his chagrin peeps out in the remark to Boswell: "Mrs. Montagu has dropped me. Now, Sir, there are people whom one would like very well to drop, but would not wish to be dropped by." The wound was never healed, and the sufferer had not always the magnanimity to conceal his pain. Baretti remarks in his 'Marginalia': "There was no good intelligence between Mrs. M. and Dr. J." Sir Joshua Reynolds happening to say that the 'Essay on Shakspeare' did her honour, Johnson's reply was: "Yes, Sir, it does *her* honour; but it would do nobody else honour." It is to be feared that the doctor's opinions thus freely expressed affected others; for sundry attacks were made in the press upon the poor lady's failings, which caused her friends much pain. Thus, Miss Burney observes, "Whatever may be Mrs. Montagu's foibles, she is free, I believe, from all vice, and as a member of society she is magnificently useful." She died in 1800; and ought to have died sooner.

Mrs. Ord is bracketed, by Wraxall, with Mrs. Montagu and Mrs. Vesey as one of the three great leaders of Blue-stocking society;

and from the constant references to her entertainments which are to be found in Walpole, D'Arblay, More, etc., it is evident that she was a star of some magnitude in the London world of fashion. Though herself, so far as we know, undistinguished by any attainments she affected literature and literary people. Johnson was a good deal at her house, especially during the year 1780. We find him going to Mrs. Ord's on a Thursday in April, and meeting her the following Sunday at Dr. Burney's. In May he meets at Mrs. Ord's, "a travelled lady of great spirit and some consciousness of her own ability." And Hannah More describes her chagrin when, during a period of court mourning for some German princess deceased in the same year, she found herself, amid a large company at Mrs. Ord's, the only guest arrayed in colours—" Even Jacobite Johnson was in deep mourning." But she afterwards writes of one of Mrs. Ord's assemblies: "Everybody was there, and in such a crowd I thought myself well off to be wedged in with Mr. Smelt, Langton, Ramsay, and Johnson. There was everything delectable in the Blue way." In fact, Mrs. Ord was one of those persons who are apparently brought into the world for the sole purpose of keeping house for society. People of fashion or distinction went to her assemblies as part of their routine, but for no other reason that we can discover. She is never reputed as having written anything, or said anything, or done anything, worth recollecting ; still, she was one of those negative people who are indispensable to town life, and was besides a good-natured lady who never did anybody harm so far as we are aware.

It is much to be feared that she who, next to Mrs. Siddons, was the most celebrated actress of her time figured amongst the Doctor's aversions. He could never abide women who were big, heavy, or vulgar ; and unless she is very much belied, Mrs. Pritchard was all three. But Mrs. Pritchard, before he knew her, had been slim and sprightly enough; and a prime favourite at Bartholomew Fair, where her singing of an arch ditty, "Sweet, if you love me, smiling turn," was a never-failing "draw." During the earlier part of her career at Drury Lane, she usually played comedy; but her versatility was remarkable, inasmuch as we find her during one and the same season filling such incongruous parts as "Lady Macbeth," and "Mrs. Doll" in Ben Jonson's 'Alchymist.' As she advanced in years and stoutness, she became more and more identified with "lofty business." Churchill praises her "Lara" in 'The Mourning Bride'; and such was her reputation as "The Queen" in 'Hamlet' that when she left the stage no substitute could for a long time be found willing to take that part. Her province was, however, considered to be on the borderland between high and low comedy; and the critics of the day proclaimed her "a perfect mistress of *familiar dramatic eloquence.*" Davies, in his 'Life of Garrick,' declares that "her delivery of dialogue, whether of humour, wit, or more sprightly conversation, was never surpassed nor perhaps equalled." But other, and we fear more reliable, authorities held that she never rose to the highest grade, even of comedy, while in tragedy she wanted grace and was prone to rant, being "too loud and profuse in her grief"; or, as Garrick put it, "too apt to blubber her sorrow." She was, further, built on so grand a scale as to be of rather overpowering presence; whilst her features were too massive to be mobile. But her gravest drawback was the indolence which increased in proportion to her bulk. It was with the greatest difficulty that she was induced to study her own parts; and, like the Fotheringay,* she never studied more. Thus, it was well known that she had never read 'Macbeth' in her life, though "Lady Macbeth" was one of her greatest parts. Johnson was aware of this, and despised her for it. Her name being once mentioned, he said: "Her playing was quite mechanical. It is wonderful how little mind she had. Sir, she had never read the tragedy of 'Macbeth' all through. She had no more thought of the play out of which her part was taken than a shoemaker thinks of the skin out of which the piece of leather of which he is making a pair of shoes is cut." The poor lady had another

* Pendennis's *innamorata* is here alluded to.

failing which militated against her social success, and rendered her particularly distasteful to the fastidious Doctor—she was incurably vulgar; though, oddly enough, as in the case of the Fotheringay, this defect was never apparent when she trod the boards. " Pritchard in common life," said the Doctor, "was a vulgar idiot; she would talk of her *gownd;* but, when she appeared upon the stage, seemed to be inspired by gentility and understanding." But probably her greatest demerit in Johnson's eyes was her being associated with the failure of his tragedy, in which she had sustained the title *rôle.* For, whether owing to her conception of that part, or to its own gruesomeness, the public would not stand her in 'Irene' at any price. Dr. Adams, who was present at the first representation of the tragedy, says: "The play went off tolerably till it came to the conclusion, when Mrs. Pritchard, the heroine of the piece, was to be strangled on the stage, and was to speak two lines with the bow-string round her neck. The audience cried out, '*Murder, murder!*' She several times attempted to speak. At last she was obliged to go off the stage alive." At subsequent representations they allowed her to be carried off and done to death behind the scenes; but, in spite of alterations, "the tragedy of 'Irene' did not please the publick"; and like many another sensitive author, Johnson declined to hold himself responsible for its want of success. Mrs. Pritchard's popularity did not increase as she grew older. Rivals disputed her supremacy; until at last, awaking to the fact that defeat might be in store for her on the field where she had hitherto triumphed, she resolved to conquer elsewhere, and transferred her weakened forces to Dublin. There, as we are pained to learn, "she electrified the Irish with disappointment." After this we hear little more of her; and she finally retired from the practice of her profession in 1768, having held the stage for over thirty years. Clive and she were always fast friends, which shows that she must have possessed an amiable temper, for Kitty was not an easy person to get on with professionally. But good nature counted for little with Johnson, where women were concerned, if not backed up by good looks and vivacity—neither of which the worthy Pritchard seems to have possessed.

The Rev. Mr. Seward, Canon Residentiary of Lichfield, was an ingenious and literary man who had published an edition of Beaumont and Fletcher, and written verses in "Dodsley's collection." At his house (where Johnson happened to be then staying) Boswell had "the pleasure of seeing his celebrated daughter, Miss Anna Seward, to whom I have since been indebted for many civilities, as well as some obliging communications concerning Johnson." Miss Seward's celebrity depended chiefly upon her gifts as a poetess which, judging from her literary remains, were not very remarkable. But Johnson, who had a kindness for the daughter of his old friend, encouraged her with unstinted praise. Once, upon her mentioning to him an epic poem, 'The Columbiade,' by Madame du Boccage, he exclaimed, "Madam, there is not anything in it equal to your description of the sea round the North Pole in your ode on the death of Captain Cook." The 'Ode' was in fact an 'Elegy,' and the description referred to is very poor stuff indeed; but when Johnson took it into his head to be gracious he did not stick at a trifle. Although Boswell was so much charmed with the lady at first, he saw reason to alter that opinion; and we afterwards find him complaining that he was obliged to receive her "obliging communications with much caution, as they were tinged with a strong prejudice against Johnson." In fact, some years after Johnson's death (1793) he roundly accused her in the 'Gentleman's Magazine' of having attacked the Doctor with malevolence. Miss Seward, who was known in London society as "the poetess of Lichfield," seems to have been of a rather captious and sarcastic turn in conversation. We find her, for instance, catching up Johnson pretty sharply once or twice during that celebrated dinner at Dilly the publisher's; and there is no doubt that she was not generally beloved. Horace Walpole has no kind word for her: "I am a little sick," he writes, "of the

Hayleys and Miss Sewards"; again, he asks: "Is it not extraordinary that Miss Seward can write no better?" And yet again, with unusual ill-humour, he complains that "Miss Seward, Miss Williams, and half-a-dozen more of those harmonious virgins, have no imagination and no novelty. Their thoughts and phrases are, like their gowns, old remnants cut and turned." There was no love lost between the two. The lady, in her turn, denounces Walpole as "One of those who weary, sicken, and disgust people whose sensibilities are strong and healthy by their eternal cant about the great *have-beens* and the little *ares.*" Altogether, a pretty quarrel.

For Mrs. Siddons Johnson had a high respect and esteem; but that lady's manner did not possess the sprightliness which to our worthy Doctor was one of the most charming qualities in a woman. Mrs. Thrale found her "leaden" in private life; and he may have done the same. When her fame was at its highest, he showed no eagerness to witness her triumphs. " But, indeed, Dr. Johnson," said Miss Monckton to him, "you *must* see Mrs. Siddons. Won't

Mrs. SIDDONS.

you see her in some fine part?" "Why, if I *must*, Madam, I have no choice," was the ungracious reply. They do not appear to have met very often; but in 1783 she paid him the compliment of visiting him at his own house. The account which Mr. Kemble gave of their interview is as follows:—

"When Mrs. Siddons came into the room there happened to be no chair ready for her, which he observing, said with a smile, 'Madam, you who so often occasion a want of seats to other people will the more easily excuse the want of one yourself.'

"Having placed himself by her, he with great good humour entered upon a consideration of the English drama; and, among other enquiries, asked her which of Shakspeare's characters she was most pleased with. Upon her answering that she thought the character of Queen Catherine in ' Henry the Eighth' the most natural, 'I think so, too, Madam,' (said he,) ' and whenever you perform it I will once more hobble out to the theatre myself.' Mrs. Siddons promised she would do herself the honour of acting his favourite part for him; but many circumstances happened to prevent the representation of 'King Henry the Eighth' during the Doctor's life."

The impression which he formed of his

visitor at the above interview shows that, had they met more frequently, a warm friendship might have been the result. For we find him afterwards writing to Mrs. Thrale: "Mrs. Siddons in her visit to me behaved with great modesty and propriety. Neither praise nor money, the two powerful corruptors of mankind, seem to have depraved her. I shall be glad to see her again." But this appears to have been their last meeting; and if we reckon the great actress among the Doctor's "friends," it is mainly because he never evinced any hostility to her.

Having already dealt with Mrs. Thrale under another aspect, we have now to consider her as one of the foremost among the literary women of her time. From familiarity, no doubt, Johnson somewhat underrated her powers; indeed, he for some reason or other affected to consider them inferior to those of her husband. "It is a great mistake," he remarked to Boswell, "to suppose that she is above him in literary attainments. She is more flippant, but he has ten times her learning; her learning is that of a schoolboy in one of the lower forms." On the other hand, we are told that she was "enchanted with Johnson's conversation for its own sake, and also had a very allowable vanity in appearing to be honoured with the attention of so celebrated a man." It must be remembered, however, that Boswell, whether from jealousy or resentment, was no friend to her, nor likely to give her credit for half she deserved. Yet even he was forced to admit that "the vivacity of Mrs. Thrale's literary talk roused him (Johnson) to cheerfulness even when they were alone." Miss Burney, who was a good deal at Streatham, describes her as having talents to create admiration, good-humour to excite love, and understanding to give entertainment . . . a sweetness of disposition that excels all her other excellences, and far from making a point of vindicating herself, she generally receives his (Johnson's) admonitions with the most respectful silence." Indeed, she made no secret of entertaining, at any rate a filial, affection for the sage. "There are many," she said to Boswell, "who admire and respect Mr. Johnson; but you and I love him"; and the philosopher himself was fain to own that "she would be the first woman in the world, could she but restrain that wicked tongue of hers." The daughter of Mr. Salusbury, a Welsh gentleman of property, she had been given an extensive education, which was considerably augmented by Johnson. In Madame d'Arblay's memoirs she alludes to certain Latin lessons which he was in the habit of giving at Streatham; and we know that in addition to that language Mrs. Thrale was acquainted with French, Italian, and Spanish, with probably some Greek thrown in. But her most useful acquirement was her extensive knowledge of general literature, which enabled her to hold her own in any company; and besides all this she had, as her biographer, Mr. Seeley, observes, "a natural talent for conversation, which she improved by constant and assiduous practice till she became one of the most famous female talkers of her time." According to Madame d'Arblay, her celebrity exceeded that of her two rivals for supremacy in the Blue-stocking world:

Dr. Johnson as a Man of Fashion. 177

"Mrs. Vesey, indeed, gentle and diffident, dreamed not of any competition; but Mrs. Montagu and Mrs. Thrale were set up as rival candidates for colloquial eminence, and each of them thought the other alone worthy to be her peer. Whenever they met, therefore, a contest for superiority ensued." But Mrs. Montagu had a quality which the other lacked—dignity; and this gave her the advantage. Wraxall confesses that Mrs. Thrale had "at least as much information, a mind as cultivated, and more wit than Mrs. Montagu; but she did not descend among men from such an eminence, and she talked much more, as well as more unguardedly, on every subject." That shrewd observer, like Johnson, hit upon her chief defect—a want of repose, of power to restrain "that wicked tongue of hers"; and she allowed this weakness to appear not only in her conversation but in her books. Alluding to Miss Burney's absurd comparison of her with Madame de Staël, the late Mr. Hayward very truly observes: " Her mind, despite her masculine acquirements, was thoroughly feminine; she had more tact than genius; more sensibility and quickness of perception than depth, comprehensiveness or continuity of thought. . . . She professed to write as she talked; but her conversation was, doubtless, better than her books, her main advantages being a well-stored memory, fertility of images, aptness of allusion and *apropos*." Indeed, while the writings she has left possess a certain value, owing to their graphic style and piquancy, it is mainly to the persons about whom she writes that they are indebted for such popularity as they possess. She had some facility as a versifier, and her "Three Warnings" which was first published in Miss Williams's 'Miscellanies,' in 1776, still figures in modern collections of verses which are not exactly poetry. Her 'Anecdotes of Dr. Johnson' and her 'Letters to and from Dr. Johnson' will always be read, though not for her own sake. But her 'Travels,' compiled during her wedding tour as Mrs. Piozzi, brought down on her a perfect storm of censure. Horace Walpole who, by the way, had no great admiration for her, is never weary of denouncing the solecisms in these unlucky volumes. "It was said," he remarks, "that Addison might have written his travels without going out of England. By the excessive vulgarities so plentiful in these volumes we might suppose the writer had never been outside the parish of St. Giles." Again, " Mrs. Piozzi is going to publish a book on English Synonyms. Methinks she had better have studied them before she stuffed her 'Travels' with so many vulgarisms." *Per contrà*, Miss Seward, who liked her and hated Walpole, praises the 'Travels' highly; but is nevertheless fain to deplore "the infinite inequality of the style." Nor did her 'Anecdotes of Dr. Johnson' find much favour at Strawberry Hill. "Signora Piozzi," writes the cynic, "is not likely to gratify her expectations of renown. Our comic performers are Boswell and Donna Piozzi. She and Boswell and their Hero are the joke of the public." All which goes to show that critics, be they never so wise, do not make nor mar the fortunes of books. Some of the censure so freely bestowed may be attributed to the lady's temporary unpopularity on account of her second marriage. Of this we have a hint in the following couplet from the 'Baviad':—

"See Thrale's gray widow with a satchel roam
And bring in pomp laborious nothings home,"

also, in Walpole's acrimonious allusion to "that ridiculous woman, Madame Piozzi." But such unpopularity is rarely more lasting than it is reasonable; and the victim was not long in regaining her social prestige. Of this we have proof in the list of those who visited her the year after her return from Italy, which includes the names of all the celebrities in town. After Piozzi's death she gave up London life and spent most of her time at Bath or Clifton, where the vivacious widow took a prominent part in any gaieties which might be going on. Indeed, she is recorded as having danced at a public assembly when in her eightieth year. She used rouge unsparingly, and made no secret of the habit; for, to do her justice, she always seemed incapable of concealment. At length, the life she had so much enjoyed was prematurely brought to a close by an accident which befell her on a journey between Clifton and Penzance. She broke her leg, and was brought back to Clifton, where the end soon came. A characteristic circumstance of her last illness was that when her medical attendant tried to cheer her with hopes of a recovery, she responded by tracing the figure of a coffin in the air with her finger, and then lay calmly down to face the inevitable. Take her all in all, she is an interesting personality not likely to be forgotten. Her features are preserved for us in Hogarth's picture of "The Lady's Last Stake," for which she tells us that she posed as the leading figure, and in numerous engravings, of which we have reproduced a couple.*

Next to Mrs. Montagu, as a leader of the Blue-stockings, ranked Mrs. Vesey. Indeed, it was thought by many people that the literary parties which the latter gave at her house in Bolton Row, and afterwards in Clarges Street, were more modish than those in Portman Square. Wraxall, for one, was certainly of opinion that "her repasts were at once more select and more delicate than those of Mrs. Montagu, though not so

* At pp. 51 and 67.

remarkable for display." Personally, she lacked the brilliant qualities of the other, being distinguished neither for her abilities nor her attainments. In fact, she was a good-natured, simple, and unobservant lady, who adored genius without having a particle of it herself. Her most remarkable mental endowment was a singularly defective memory. It is related of her that she often forgot her own name; and that when upon a certain occasion someone thought it necessary to remind her that Mr. Vesey was her *second* husband, the oblivious lady exclaimed in surprise, "Bless me, my dear! I had forgotten all about it." She was much older than Mrs. Montagu, did not possess the latter's love of finery; nor, though very wealthy, did she seek to rival the magnificence of that lady's establishment. Miss Burney describes her as having "the most wrinkled, sallow, time-beaten face I ever saw. She is an exceedingly well-bred woman, and of agreeable manners; but all her name in the world must, I think, have been acquired by her dexterity and skill in selecting parties and by her dexterity in rendering them easy

Dr. Johnson as a Man of Fashion. 183

with one another." Everybody seems to have liked, and quizzed, the old lady. Horace Walpole, who used to call her parties "Babels," because (as he asserted) the guests were unintelligible to each other, has a kind word for the good old woman. "A warm heart feels itself ready to do more than is possible for those it loves," he wrote * at a time when she lay ill; "I am sure our poor friend in Clarges Street would subscribe to this last sentence. What English heart ever excelled hers?" Johnson was frequently at her assemblies, as we learn from Wraxall, Walpole, and Hannah More; but does not appear to have given his hostess much of his conversation—in fact, nobody seems to have troubled the poor lady with any profound remarks. Sometimes she rather bored them by her anxiety to put everybody at their ease. "Mrs. Vesey," said a noble lord at Mrs. Cholmondeley's, "is vastly agreeable, but her fear of ceremony is really troublesome; for her eagerness to break a circle is such that she insists upon everybody's sitting with their backs one to another;

* To Hannah More.

that is, the chairs are drawn into little parties of three together, in a confused manner, all over the room." * It was further complained that her assemblies were sometimes "too full to be very pleasant."

Miss Helen Maria Williams was one of the many people about whom Boswell saw fit to change his first-conceived opinion. We are introduced to her with a flourish :—

"He (Johnson) dined that day at Mr. Hoole's, and Miss Helen Maria Williams being expected in the evening, Mr. Hoole put into his hands her beautiful 'Ode on the Peace.' Johnson read it over, and when this amiable, elegant and accomplished young lady was presented to him, he took her by the hand in the most courteous manner, and repeated the finest stanza of her poem ; this was the most delicate and pleasing compliment he could pay."

In a subsequent edition we have the reverse of the medal :—

"In the first edition of my work the epithet *amiable* was given. I was sorry to be obliged to

* Madame d'Arblay.

Dr. Johnson as a Man of Fashion. 185

strike it out; but I could not in justice suffer it to remain, after this young lady had not only written in favour of the savage anarchy with which France has been visited; but had (as I have been informed on good authority) walked, without horror, over the ground of the Tuileries when it was strewed with the naked bodies of the faithful Swiss guards. . . . From Dr. Johnson she could now expect not endearment but repulsion."

It was to this young lady that Johnson paid one of those elaborate compliments already noticed. He had asked her to sit down by him, which she did; and upon her inquiring how he was, he answered, "I am very ill indeed, Madam. I am very ill when you are near me; what should I be were you at a distance?" From which we may presume that she was not wanting in attractions; and indeed Rogers, the poet, describes her as "a very fascinating person." At the time of this interview she was about twenty-two years of age, having been born in 1762; but had already made herself remarkable as an exponent of the advanced "French ideas of liberty" so much in favour with the New Woman of that day. Her proclivities led her to Paris, where she became at first a warm supporter of the Revolution; but having afterwards espoused the cause of the Girondists, was imprisoned in the Temple, and only released at the fall of Robespierre. Undaunted by this experience, she returned to Paris in 1796, and died there in 1827, having, in her later political writings, shown herself a friend of the Bourbons and an enemy of the Revolution. As an author of poems, plays, novels, books of travel, and translations, she had some repute; but is chiefly known as a voluminous writer on political subjects. She was also a contributor to the Press, and for some years wrote that part of 'The Annual Register' which relates to France.

IV.
DR. JOHNSON ON DRESS AND DEPORTMENT.

THERE is no graver misconception in the popular estimate of Johnson than that which supposes him indifferent to externals. Few men had a keener eye for what was becoming or unsuitable in dress, or a nicer discrimination regarding propriety of demeanour. Ladies whose toilettes had passed triumphantly the ordeals imposed by their own sex quailed before his searching gaze; and the frisky matron, like the thoughtless hoyden, curbed her eccentricities when in his presence. The Doctor had loftier ideas than most people on the subject of dress. It should, in his opinion, form an index to the real or representative character of the person wearing it. It should be positive, as a clue to social position; if plain, "plain to severity"; if fine, "the finer the better." He himself even deigned to furnish upon at least one occasion a remarkable exemplification of his views on the subject. It was while his tragedy of 'Irene' was being performed, when he considered that, as a dramatic author, his dress should be more gay than what he ordinarily wore; and gay it accordingly was, with a gaiety beyond the dreams of authorship. We are told that during the representation "he appeared behind the scenes, and even in one of the side-boxes, in a scarlet waistcoat, with rich gold lace, and a gold-laced hat." The effect was prodigious; and the memory of that display was long cherished by the sage. Years afterwards, he remarked musingly to Boswell: "Sir, I once had a very rich laced-waistcoat, which I wore the first night of my tragedy." And though it must be owned that, as a rule, he was somewhat careless of his attire in private life, and occasionally permitted himself to relapse into a certain disregard of those observances which are practised in ordinary society, he rarely countenanced a similar laxity in others—never in women. No milliner of Bond Street could be more critical to detect the displacement of a ribbon, the want of modishness in a cap, or inharmonious colouring in a dress. He had theories on such matters, which were never unsupported by argument and always nicely graduated to the several ages, stages, functions, and ceremonies of life; but having more especial reference to distinctions of social position and the due enhancement of female charms. To explain what these doctrines were, we cannot do better than furnish the reader with an excerpt from the journal of a lady who had very good reason for being well-informed on the point:—

"Even dress itself, when it resembled that of the vulgar, offended him exceedingly; and when he had condemned me many times for not adorning my children with more show than I thought useful or elegant, I presented a little girl to him who came a-visiting one evening covered with shining ornaments, to see if he would approve of the appearance she made. When they were gone home, 'Well, Sir,' said I, 'how did you like little Miss? I hope she was *fine* enough.' 'It was the finery of a beggar,' said he, 'and you know it was; she looked like a native of Cow Lane dressed up to be carried to Bartholomew Fair.'

"His reprimand to another lady for crossing her little child's hand kerchief before, and by that operation dragging down its head oddly and unintentionally, was on the same principle. 'It is the beggar's fear of cold,' said he, 'that prevails over such parents, and so they pull the poor thing's head down, and give it the look of a baby that plays about Westminster Bridge, while the mother sits shivering in a *niche*.'

"I commended a young lady for her beauty and pretty behaviour one day, however, to whom I thought no objection could have been made. 'I saw her,' says Dr. Johnson, 'take a pair of scissors in her left hand though; and for all her father is now become a nobleman, and as you say, excessively rich, I should, were I a youth of quality ten years hence, hesitate between a girl so neglected, and a *negro*.'

"It was indeed astonishing how he *could* remark such minuteness with a sight so miserably imperfect; but no accidental position of a ribband escaped him, so nice was his observation, and so rigorous his demands of propriety. When I went with him to Lichfield, and came downstairs to breakfast at the inn, my dress did not please him, and he made me alter it entirely before he would stir a step with us about the town, saying most satirical things concerning the appearance I made in a riding-habit, and adding, ''Tis very strange that such eyes as yours cannot discern propriety of dress. If I had a sight only half as good, I think I should see to the centre.'

"My compliances, however, were of little worth. What really surprised me was the victory he gained over a lady little accustomed to contradiction, who had dressed herself for church at Streatham one Sunday morning in a manner he did not approve, and to whom he said such sharp and pungent things concerning her hat, her gown, etc., that she hastened to change them, and returning quite another figure, received his applause, and thanked him for his reproofs, much to the amusement of her husband, who could scarcely believe his own ears.

"Another lady, whose accomplishments he never denied, came to our house one day covered with diamonds, feathers, etc., and he did not seem inclined to chat with her as usual. I asked him why, when the company was gone. 'Why, her head looked so like that of a woman who shows puppets,' said he, 'and her voice so confirmed the fancy, that I could not bear her to-day. When she wears a large cap I can talk to her.'

"When the ladies wore lace trimmings to their clothes he expressed his contempt of the reigning fashion in these terms: 'A Brussels trimming is like bread sauce,' said he; 'it takes away the glow of colour from the gown, and gives you nothing instead of it. But sauce was invented to heighten the flavour of our food, and trimming is an ornament to the manteau or it is nothing. Learn,' said he, 'that there is propriety or impropriety in everything how slight soever, and get at the general principles of dress and of behaviour; if you transgress them you will at least know that they are not observed.'" *

Miss Burney, also, was not long in discovering what a martinet our Doctor was in respect of ladies' attire. Almost immediately after her arrival at Streatham she has occasion to record in her diary:—

" It seems that he always speaks his mind concerning the dress of ladies, and all ladies who are here obey his injunctions implicitly, and alter whatever he disapproves. This is a part of his character that much surprises me; but notwithstanding he is sometimes so absent, and always so near-sighted, he scrutinises into every part of almost everybody's appearance. They tell me of a Miss Brown, who often visits here, and who has a slovenly way of dressing: 'And when she comes down in a morning,' says Mrs. Thrale, 'her hair will be all loose, and her cap half off; and then Dr. Johnson, who sees something is wrong, and does not know where

* 'Piozzi Anecdotes.'

the fault is, concludes it is in the cap, and says, "My dear, what do you wear such a vile cap for?" "I'll change it, Sir," cries the poor girl, "if you don't like it." "Ay, do," he says; and away runs poor Miss Brown; but when she gets on another, it's the same thing, for the cap has nothing to do with the fault. And then she wonders that Dr. Johnson should not like the cap, for she thinks it very pretty. And so on with her gown, which he also makes her change; but if the poor girl were to change all through her wardrobe, unless she could put her things on better, he would still find fault.'"

Miss Burney's own caps soon became the subject of observation. Looking earnestly at her one day, he broke out with :—

"' It's very handsome!'

"' What, Sir?' cried I, amazed.

"' Why, your cap : I have looked at it some time, and I like it much. It has not that vile bandeau across it, which I have so often cursed.'"

No age or infirmity seems to have been exempt from the Doctor's iron rule as regarded dress. Even poor old Mrs. Burney (Fanny's mother) was sent off to change her gown because it did not please him ; and when she had done so, was told that she "should not wear a black hat and cloak in summer." Thinking to propitiate him, she informed him one day that she had "got her old white cloak scoured on purpose to oblige him. 'Scoured!' says he, 'Aye, have you, Madam?' So he seesawed,—for he could not for shame find fault, but he did not seem to like the *scouring.*" Boswell has already told us how when Mrs. Thrale appeared before him in a dark gown he rebuked her with, "You little creatures should never wear those sort of clothes; they are unsuitable in every way. What! have not all insects gay colours?" From which, and poor Mrs. Burney's experience, we may gather that the Doctor's fancy was for bright colours. But be that as it may, he was allowed by the ladies to possess "a most exact and elegant taste in dress"—another proof, if proof were wanting, of what a wonderful man he was. Nor, upon the whole, can it be disputed that there is a grain of reason in his severe remark, "The truth is, women, take them in general, have no idea of grace. Fashion is all they think of."

His doctrine of deportment, if equally

precise, was less elaborate. It might, perhaps, be summed up in the single maxim that women were bound above all things to be *genteel*; but the examples which he gives of its application are not always happy. "Were a woman sitting in company to put out her legs before her, as most men do, we should be tempted to kick them in," is stating a most improbable hypothetical case in a decidedly repulsive form. It was, however, uttered in the heat of conversation, which may account for its startling brusqueness; and in treating of so delicate a subject we prefer to ascertain the views which he expressed with deliberation in the calmer atmosphere of his private work-room. The *Adventurer*, the *Idler*, and the *Rambler* contain essays upon the subject of female carriage and deportment which, although not always written by Johnson himself, bear the stamp of his approval; and from these we are able to gather a few rules of practice. Thus we find young ladies specially warned against making eyes when at church, "by which they vainly hope to attract admirers"; for, "women are always most observed when they seem themselves least to observe, or to lay themselves out for observation!" At church, the lady's "eyes should be her own, her ears the preacher's." A woman should be, in her conversation, habitually cautious; she should speak well of everybody, wear an unruffled demeanour at all times, and learn to smile "not so much by sensation as by practice." She should be tolerant when gentlemen make jokes, and "should not object to hearing the same joke over again." She should be "an enemy to nothing but ill-nature and pride"; she should never hate anybody—at least *openly*, and should take all opportunities to show how easily she can forgive. Whilst not neglecting the higher accomplishments, she should manage to be generally "found employed in domestic duties"; and she should always bear in mind that "modesty and diffidence, gentleness and meekness are looked upon as the appropriate virtues and characteristic graces of the sex." If a young gentleman of suitable position gives a young lady reason to suspect that he has "intentions," she must, even if favourably disposed to the aspirant, keep him at arm's length, so to speak, until their respective parents have sanctioned his advances. Then indeed she may accord him an interview; but should be "modestly reserved," during their first meeting, at any rate, and confine herself to generalities—such as owning that she is "obliged to him for his good opinion"; though she may "tacitly allow of his future visits." The rest is all plain-sailing, and needs no rules.

So far, our grave philosopher has confined himself to the guidance of those fair creatures who, having lately crossed the threshold of womanhood, are entering upon the serious business of life; but he is not forgetful of those dames who, having played their part in the arena, should be thinking of laying down those weapons with which they have been used to conquer. Towards these, it must be confessed, he does not comport himself with much suavity—gives them a good moral "shaking," and puts his counsel with distressing plainness. In scathing language he deplores the fact that "ladies are desirous to hide from themselves the advances of age, and endeavour to supply the sprightliness and bloom of youth by artificial beauty and forced vivacity. They hope to inflame the heart by charms which have lost their fire, or melt it by languor which is no longer delicate. They play now the airs which pleased at a time when they were expected only to please, and forget that airs in time should give place to virtues. They continue to trifle because they could once trifle agreeably, till those who shared their early pleasures are withdrawn to more serious engagement; and are scarcely awakened from their dream of perpetual youth, but by the scorn of those whom they endeavour to rival." Bitter words, indeed.*

Johnson's theory of deportment was in reality very simple. It might be summed up in one word—*Restraint.* "There are ten genteel women," he observes, "for one genteel

* They may not, however, have been Johnson's, though in his capacity of responsible editor he must have adopted them. Mrs. Carter is suspected of having written the greater part of the essay from which they are taken, No. 34 of the *Rambler.*

man, because they are more restrained. A man without some degree of restraint is insufferable; but we are all less restrained than women." So that, after all, his doctrine of good manners is pretty much that which has obtained in all ages and among all peoples. For what is that which has ever constituted the well-bred man, or woman, but the power of repressing natural impulse, whether to undue eagerness, resentment, curiosity, or shyness? The difference between well-mannered persons of a thousand years ago and of to-day, between the Frenchman, the Englishman, and the Antipodean, is merely one of degree— each practising his own particular form of self-repression. Of course, as Johnson points out, "an elegant manner and easiness of behaviour are acquired gradually and imperceptibly." The natural man or woman, no matter what has been said to the contrary, cannot, owing to the very nature of humanity, be other than a savage; indeed, the very use of the term "good *breeding*," to denote good manners, implies the use of disciplinary reform. In youth the restraint is imposed by others, in later life we practise it ourselves; and in proportion to the extent to which it becomes habitual are we more or less well-mannered. Imitation is of course a great help; and, consequently, persons accustomed to the society of the upper classes, who inherit the traditions of a perfected system of restraint, are likely to be better-mannered than those who have not the same opportunities. But it all comes to the same thing in the end; and however we may differ in our appreciation of the special rules which Johnson lays down for the practice of what he calls "gentility," there can, we fancy, be little doubt that the principle which lies at the bottom of them is perfectly sound. Some of these rules are so manifestly out of date as to have become ridiculous; but as society was constituted in his day they recommended themselves to the great majority of mothers and daughters. We now smile at his deprecation of portrait-painting as an employment for ladies, because "Public practice of any art, and staring in men's faces, is very indelicate in a female." Things have changed a good deal in the art-world since that view prevailed, but people did not see anything absurd in it then. And what a different order of life from ours do the following utterances point to!—

"'Stocks for the men, a ducking-stool for women, and a pound for beasts. If we require more perfection from women than from ourselves it is doing them honour. And women have not the same temptations that we have; they may always live in virtuous company; men must mix in the world indiscriminately. If a woman has no inclination to do what is wrong, being secured from it is no restraint to her. I am at liberty to walk into the Thames; but if I were to try it, my friends would restrain me in Bedlam, and I should be obliged to them.' MRS. KNOWLES: 'Still, Doctor, I cannot help thinking it a hardship that more indulgence is allowed to men than to women. It gives a superiority to men, to which I do not see how they are entitled.' JOHNSON: 'It is plain, Madam, one or other must have the superiority. As Shakspeare says, "If two men ride on horseback, one must ride behind."' DILLY: 'I suppose, Sir, Mrs. Knowles would have them to ride in panniers, one on each side.' JOHNSON: 'Then, Sir, the horse would throw them both.' MRS. KNOWLES: 'Well, I hope that in another world the sexes will be equal.' BOSWELL: 'That is being too ambitious, Madam.

We might as well desire to be equal with the angels. We shall all, I hope, be happy in a future state, but we must not expect to be all happy in the same degree. A worthy carman will get to heaven as well as Sir Isaac Newton. Yet, though equally good, they will not have the same degrees of happiness.' JOHNSON: '*Probably not.*'" *

To pursue this subject might lead to painful misconceptions of the Doctor's well-merited character for devotion to the sex. In the heat of argument he was often led by the instigation of others to say more than he meant; and this may have been the case here. At all events, ladies should give him the benefit of the doubt.

* Boswell's account of a dinner at Mr. Dilly's.

V.
Dr. Johnson on Marriage and the Relations of the Sexes.

As a man of the world, no less than a philosopher, Dr. Johnson took a keen interest in all social questions; but from his personal experience of love and marriage he was peculiarly fitted to discuss those relations upon which family life and the continuance of our species are founded. The topic was one which would naturally crop up during those miscellaneous conversations in which he was prone to engage with people of both sexes; and the views to which he gave expression on such occasions are those of a man who had long and carefully studied the subject in all its bearings. With regard to love, it is somewhat disappointing to find that he, whose experience thereof had been so wide and profound, treats it as a matter of comparatively little importance. " Of the passion of love, he remarked that its violence and ill effects were much exaggerated ; for who has known any real sufferings on that head, more than from the exorbitancy of any other passion?" In this we fancy the verdict of the world will be against him ; as also in the extremely latitudinarian view expressed in the following passage :—

BOSWELL : " Pray, Sir, do you not suppose that there are fifty women in the world with any one of whom a man may be as happy as with any one woman in particular?" JOHNSON : " Aye, fifty thousand." BOSWELL : " Then, Sir, you are not of opinion with some who imagine that certain men and certain women are made for each other, and that they cannot be happy if they miss their counterparts?" JOHNSON : "To be sure not, Sir. I believe marriages would in general be as happy, and often more so, if they were all made by the Lord Chancellor upon a due consideration of characters and circumstances, without the parties having any choice in the matter."

We find him repeating the same opinion more than once. Mrs. Thrale, for instance, heard him say : " If you shut up any man with any woman, so as to make them derive their

whole pleasure from each other, they would inevitably fall in love, as it is called, with each other; but at six months' end, if you would throw them both into public life, where they might change partners at pleasure, each would soon forget that fondness which mutual dependence and the paucity of general amusement alone had caused, and each would separately feel delighted by their release."*
But for all that, it is doubtful that in his heart of hearts there did not exist an ideal of love very similar to that of ordinary folk. Now and then the weakness betrays itself:—

" A lady at my house," writes Mrs. Thrale, " said she would make him talk about love; and took her measures accordingly, deriding the novels of the day because they treated about love. ' It is not,' replied our philosopher, ' because they treat, as you call it, about love, but because they treat of nothing, that they are despicable. We must not ridicule a passion which he who never felt never was happy, and he who laughs at never deserves to feel—a passion which has caused the change of empires and the loss of worlds—a passion which has inspired heroism and subdued avarice.' He thought

* ' Piozzi Anecdotes.'

he had already said too much, 'A passion, in short,' added he with an altered tone, 'that consumes me away for pretty Fanny here; and she is very cruel,' speaking of another lady in the room."

When Baretti announced to his friend that he was engaged to be married, the counsel given, if not exactly encouraging, is sound; and, as the Doctor takes his standpoint by that borderland of affection whereupon the love of courtship and that of marriage alike trench, we may append it:—

"Of your love I know not the propriety, nor can estimate the power; but in love, as in every other passion of which hope is the essence, we ought always to remember the uncertainty of events. There is, indeed, nothing which so much seduces reason from vigilance as the thought of passing life with an amiable woman; and if all would happen that a lover fancies, I know not what other terrestrial happiness would deserve pursuit. But love and marriage are different states. Those who are to suffer the evils together, and to suffer often for the sake of one another, soon lose that tenderness of look, and that benevolence of mind, which arose from the participation of unmingled pleasure and successive amusement. A woman, we are sure, will not be always fair; we are not sure she will be always virtuous; and man cannot retain through life that respect and assiduity by which he pleases for a day or for a month. I do not, however, pretend to have discovered that life has anything more to be desired than a prudent and virtuous marriage; and therefore know not what counsel to give you."

But after all, the sage confesses that love has more to do with the matter than prudence. " It is not, Sir, from reason and prudence that people marry, but from inclination. A man is poor; he thinks, ' I cannot be worse, and so I'll e'en take Peggy.'" Yet, fearful lest his friend should build too much on hope, he adds, " Now that you are going to marry, do not expect more from life than life will afford. You may often find yourself out of humour, and you may often think your wife not studious enough to please you; and yet you may have reason to consider yourself as upon the whole very happily married." It is, however, curious to remark that, while relegating love to a somewhat subordinate position among the antecedents necessary to a happy union, he draws a marked distinction between marriages in which it plays the principal part and those which are usually termed "marriages of convenience." "Our marriage service," he observed, "is too refined. It is calculated only for the best kind of marriages; whereas, we should have a form for matches of convenience, of which there are many"; though, strange to say, his notions on this point were, for a strict Churchman, somewhat lax. "He agreed with me," says Boswell, "that there was no absolute necessity for having the marriage ceremony performed by a regular clergyman, for this was not commanded in scripture."

Marriages of State were placed by him, as by most other people, in a distinct class. He disapproved of the Royal Marriage Bill, "because (said he) I would not have the people think that the validity of marriage depends on the will of man, or that the right of a king depends on the will of man. I should not have been against making the marriage of any of the royal family, without the approbation of King and Parliament, highly criminal." Tall doctrine this!

The abstract question, whether the state of marriage was *natural* to man, having been started in conversation at General Paoli's, Johnson answered it with a decided negative :—

"'Sir, it is so far from being natural for a man and woman to live in a state of marriage, that we find all the motives which they have for remaining in that connection, and the restraints which civilised society imposes to prevent separation, are hardly sufficient to keep them together.' The General said that, in a state of nature, a man and woman uniting together would form a strong and constant affection by the mutual pleasure each would receive; and that the same causes of dissension would not arise between them as between husband and wife in a civilised state. JOHNSON : ' Sir, they would have dissensions enough, though of another kind. One would choose to go a-hunting in this wood, the other in that; one would choose to go a-fishing in this lake, the other in that; or, perhaps, one would choose to go a-hunting, when the other would choose to go a-fishing; and so they would part. Besides, Sir, a savage man and a savage woman meet by chance; and when the man sees another woman that pleases him better, he will leave the first.'"

We confess, though with diffidence, that the Doctor's reasoning on this occasion does not seem to us absolutely convincing. He leaves out of sight a factor which would in our opinion have considerable influence in precluding the "dissensions" referred to; namely, the superior physical strength of the "natural man." If the lady expressed her resolve to go a-fishing when her lord's inclination was to go a-hunting, we are disposed to think that the "natural man" would put a stop to the discussion with a club, and that the two would live happily ever after, or at least so long as the "natural man" chose. Such is the *modus operandi* of the natural man as we have ever found him.

When brought into actual contact with matrimonial happiness, no one could be more frank than he in recognising the advantages of the married state. While he and Boswell were at Lichfield, it seems that "We all met at dinner at Mr. Lloyd's, where we were entertained with great hospitality. Mr. and Mrs. Lloyd had been married in the same year with their Majesties, and, like them, had been blessed with a numerous family of fine children, their numbers being exactly the same. Johnson said, 'Marriage is the best state for man in general; and every man is a worse man, in proportion as he is unfit for the married state.'" Nay, he was further of the opinion that man gains more than woman by their union; for we find him remarking afterwards to Boswell, "Marriage, Sir, is much more necessary to a man than to a woman; for he is much less able to supply himself with domestick comforts. You will recollect my saying to some ladies the other day that I had often wondered why young women should marry, as they have so much more freedom, and so much more attention paid to them, while unmarried than when married." At this end of the nineteenth century it is doubtful whether the latter statement should pass unquestioned; but in Johnson's day it might have been true enough—the "frisky matron" not having appeared as yet.

He was very strong upon the subject of *mésalliance*. "It is commonly a weak man who marries for love," he observed. On which, says Boswell, "we then talked of marrying women of fortune; and I mentioned a common remark, that a man may be, upon the whole, richer by marrying a woman with a very small portion, because a woman of fortune will be proportionately expensive; whereas a woman who brings none will be very moderate in expenses." Johnson laughed the crude notion to scorn, as it deserved; saying, "Depend upon it, Sir, this is not true. A woman of fortune, being used to the handling of money, spends it judiciously; but a woman who gets the command of money for the first time upon her marriage has such a gust in spending it that she throws it away with great profusion."

Nor upon the cognate subject of marrying for beauty does his trumpet give a more uncertain sound. "I mentioned," says Boswell, "a friend of mine having resolved never to marry a pretty woman. JOHNSON: 'Sir, it is a very foolish resolution to resolve not to marry a pretty woman. Beauty is of itself very estimable. No, Sir, I would prefer a pretty woman, unless there are objections to her. A pretty woman may be foolish;

a pretty woman may be wicked; a pretty woman may not like me. But there is no such danger in marrying a pretty woman as is apprehended; she will not be persecuted if she does not invite persecution. A pretty woman, if she has a mind to be wicked, can find a readier way than another; and that is all.'"

It is very satisfactory to find Johnson maintaining stoutly on one occasion that a father had no right to control the inclinations of his daughters in marriage, though we are fain to confess that there are utterances of his not easily to be reconciled with this amiable doctrine.

To any severance of the marriage obligation, save by the hand of death, or by the law, he was strenuously opposed. Boswell having repeated to him "an argument of a lady of my acquaintance, who maintained that her husband, having been guilty of numberless infidelities, released her from conjugal obligations, because they were reciprocal," Johnson burst out with, "This is miserable stuff, Sir. To the contract of marriage, besides the man and wife, there is a third party—society ; and, if it be considered as a vow—God : and therefore it cannot be dissolved by their consent alone. Laws are not made for particular cases, but for mankind in general. A woman may be unhappy with her husband ; but she cannot be freed from him without the approbation of the civil and ecclesiastical power." And again, "Between a man and his Maker it is a different question ; but between a man and his wife a husband's infidelity is nothing. They are connected by children, by fortune, by serious considerations of community. Wise married women don't trouble themselves about infidelity in their husbands." Touching severance by death, he was only too well qualified to speak. " He that outlives a wife whom he has long loved sees himself disjoined from the only mind that has the same hopes and fears and interests ; from the only companion with whom he has shared much good or evil ; and with whom he could set his mind at liberty, to retrace the past, or anticipate the future. The continuity of being is lacerated ; the settled course of sentiment and action is stopped; and life stands suspended and motionless, till it is driven by external causes into a new channel. But the time of suspense is dreadful."*

With regard to the rule for choosing a wife, Johnson treated it as simply an application of the general principle governing every description of choice. As Mrs. Thrale informs us:—

" The general and constant advice he gave when consulted about the choice of a wife, or profession, or whatever influences a man's particular and immediate happiness, was always to reject no positive good from fears of its contrary consequences. 'Do not,' said he, 'forbear to marry a beautiful woman if you can find such, out of a fancy that she will be less constant than an ugly one ; nor condemn yourself to the society of coarseness and vulgarity for fear of the expenses or other dangers of elegance and personal charms, which have been always acknowledged as a positive good, and for the want of which there should be always given some weighty compensation. I have, however, seen some prudent fellows who forbore to connect themselves with beauty lest coquetry should be near,

* Letter from Johnson to Dr. Lawrance, Jan. 20, 1780.

and with wit or birth lest insolence should lurk behind them, till they have been forced by their discretion to linger life away in tasteless stupidity, and choose to count the moments by remembrance of pain instead of enjoyment of pleasure.'"

He did not think that difference in religion or politics counted for much in the selection of a partner. "A Tory," said he, "will marry into a Whig family, and a Whig into a Tory family, without any reluctance. But indeed in a matter of much more concern than political tenets, and that is religion, men and women do not concern themselves much about difference of opinion." .

His attention having been once drawn to a circumstance which, we have no doubt, is of the rarest occurrence—that of a wife making a little private fund for herself out of her husband's earnings, he let his opinion of the lady's conduct be seen with tolerable distinctness. We may give the incident as described by Boswell :—

" The wife of one of his acquaintance had fraudulently made a purse to herself out of her husband's fortune. Feeling a proper compunction in her last moments, she confessed how much she had secreted; but before she could tell where it was placed she was seized with a convulsive fit and expired. Her husband said he was more hurt by her want of confidence in him than by the loss of his money. ' I told him (said Johnson) that he should console himself; for *perhaps* the money might be *found*, and he was *sure* that his wife was *gone.*' "

Though averse from completely ostracising foolish people who marry beneath them, Johnson was of opinion that the offence was one which should not be passed over. Boswell narrates, with a transparent side-allusion, how

" A young lady, who had married a man much her inferior in rank, being mentioned, a question arose how a woman's relations should behave to her in such a situation; and, while I recapitulate the debate, and recollect what has since happened, I cannot but be struck in a manner that delicacy forbids me to express. While I contended that she ought to be treated with an inflexible steadiness of displeasure, Mrs. Thrale was all for mildness and forgiveness, and, according to the vulgar phrase, making the best of a bad bargain. JOHNSON: ' Madam, we must distinguish. Were I a man of rank, I would not let a daughter starve who had made a mean marriage; but, having voluntarily degraded herself from the station which she was originally entitled to hold, I would support her only in that which she herself has chosen; and would not put her on a level with my other daughters. You are to consider, Madam, that it is our duty to maintain the subordination of civilised society; and when there is a gross and shameful deviation from rank, it should be punished so as to deter others from the same perversion.'"

He was ready enough to give advice to all aspirants after matrimony who chose to consult him, but required that the subject should be approached in a becoming manner, with due regard to its own importance and the dignity of the person consulted. One day when, at Streatham, he happened to be musing over the fire, Mrs. Thrale mentions that a young gentleman who was present "called to him suddenly, and I suppose he thought disrespectfully, in these words: 'Dr. Johnson, would you advise me to marry?' 'I would advise no man to marry, Sir,' returns for answer, in a very angry tone, Dr. Johnson, 'who is not likely to propagate understanding'; and so left the room." But it is fair to add that, when his indignation at the ill-timed intrusion upon his meditation had cooled, he returned, and spoke very nicely to the astonished lad whom he had treated so unceremoniously.

His views with regard to second marriages are not always easy to reconcile; but we may gather from the following that, upon the whole, he did not disapprove of them :—

"When," says Boswell, "I censured a gentleman of my acquaintance for marrying a second time, as it showed a disregard of his first wife, he said, 'Not at all, Sir. On the contrary, were he not to marry again, it might be concluded that his first wife had given him a disgust to marriage; but by taking a second wife he pays the highest compliment to the first, by showing that she made him so happy as a married man that he wishes to be so a second time.' So ingenious a turn did he give to this delicate question. And yet, on another occasion, he owned that he once had almost asked a promise from Mrs. Johnson that she would not marry again, but had checked himself. . . . I presume that her having been married before had, at times, given him some uneasiness; for I remember his observing upon the marriage of one of our common friends, 'He has done a very foolish thing,

Sir: he has married a widow, when he might have had a maid.'"

It was probably the opportunity for saying a good thing which alone prompted him to remark, when informed that a gentleman who, having been very unhappy in marriage, had re-married immediately after his wife died : " Sir, it was the triumph of hope over experience."

Of long-deferred marriage he strongly disapproved, observing that "more was lost, in point of time, than compensated for by any possible advantages. Even ill-assorted marriages were preferable to cheerless celibacy."

Well-grounded religious principles he considered an absolute necessity in a wife. Mr. Seward once heard him say that " a man has a very bad chance for happiness in the married state, unless he marries a woman of very strong and fixed principles of religion."

Of that rarely-forgiven offence against society, a man's marrying his own servant, he was tolerant, at least in one case. "I have been (said he) to see my old friend, Sack. Parker ; I find he has married his maid ; he has done right. She had lived with him many years in great confidence, and they had mingled minds ; I do not think he could have found any wife that would have made him so happy." And there is an amusing incident, connected with his defence of a brother author who had married "a printer's devil" (there were female devils in those days), which may be found in the description given by Boswell of a dinner at Mrs. Garrick's, on Friday, April 20, 1781.

With regard to those occasional misunderstandings which are not unknown, we believe, in married life, Johnson's habit was to side with the husband ; "whom," he said, "the woman had probably provoked so often, she scarce knew when or how she had disobliged him first."

"'Women,' says Dr. Johnson, 'give great offence by a contemptuous spirit of non-compliance on petty occasions. The man calls his wife to walk with him in the shade, and she feels a strange desire just at that moment to sit in the sun; he offers to read her a play, or sing her a song, and she calls the children in to disturb them, or advises him to seize that opportunity of settling the family accounts.

Twenty such tricks will the faithfullest wife in the world not refuse to play, and then look astonished when the fellow fetches in a mistress. Boarding-schools were established,' continued he, 'for the conjugal quiet of the parents. The two partners cannot agree which child to fondle, nor how to fondle them, so they put the young ones to school, and remove the cause of contention. The little girl pokes her head; the mother reproves her sharply. "Do not mind your mamma," says the father, "but do your own way." The mother complains to me of this. "Madam," said I, "your husband is right all the while; he is with you but two hours of the day, perhaps, and then you tease him by making the child cry. Are not ten hours enough for tuition? and are the hours of pleasure so frequent in life that when a man gets a couple of quiet ones to spend in familiar chat with his wife, they must be poisoned by petty mortifications?"'*

Which, of course, is very sound advice, as most advice is which recommends one party to a dispute to give way for the sake of peace and quietness; but a little hard on the wife who, being chiefly responsible for the bringing up of the children, is naturally mortified at seeing her instructions set at naught by their father. Nor is the Doctor's recommendation

* 'Piozzi Anecdotes.'

to "Put Missy to school," as the best remedy, one likely to satisfy an affectionate mother who does not care to part with "Missy," and yet would like that young person to be trained in the way that she should go. Still, we have nothing but praise for the general principle which pervades the entire passage—namely, that the husband's claims are paramount, and that every other consideration should be postponed to that of his perfect comfort and enjoyment. Heaven forbid that the contrary doctrine should ever prevail!

He observed with great justice that a man of sense and education should look for a suitable companion in a wife. "It was a miserable thing when the conversation could only be such as, whether the mutton should be boiled or roasted, and probably a dispute about that." And when a gentleman talked to him of a lady whom he greatly admired, but was afraid of her superiority of talents, "Sir (said he), you need not be afraid; marry her. Before a year goes round you will find that reason much smaller, and that wit not so bright."

VI.

DR. JOHNSON AS A KNIGHT-ERRANT.

So long as Johnson lived, the age of chivalry survived; it was not till eight years afterwards that Burke had to proclaim its death. No more duteous knight than he ever went forth of old to do his devoir on behalf of distressed womanhood. What if he wore no shining mail, but a snuff-coloured suit; or no nodding plume, save the scorched top-knot of his ill-fitting wig? What if he bestrode no prancing destrier, carried no blazoned shield, and bore no weapon but his stout oaken staff, was he the worse knight for that? I trow not; the courage of the man was as high, his spirit as pure, and his heart as tender as with any Galahad or Percivale of them all. But it was not in quest of young and beauteous dames that our good knight sallied forth. *His* devoir lay among the aged, the poor, the unlovely, the peevish—those shrivelled relics of long-past youth and charm from whom other men turned away unheeding or with scorn. For them he battled with those grisly caitiffs, want, sickness, and despair; he raised their poor old frames in his powerful arms, and bore them off—not to an enchanted castle, but to his rather disenchanting quarters in Bolt Court. There he tended the wounds and bruises they had received, nourished them at his simple board, and shared with them his meagre purse. More wondrous still, he, so fierce to others, so loud and masterful, so impatient of contradiction, so intolerant of human foibles, was never harsh or ungentle to these ill-fortuned ladies. If instead of thanks they gave him reproaches; instead of comfortable words, peevish complaints; instead of domestic carefulness, neglect and sorry fare, he bore it all without a murmur. Sorrow and suffering were to him in the place of loveliness and charm; and he bent his proud head before them, all meekness and submission. His voice grew soft when he addressed them; his manner tender and respectful, like those knights of the Round Table, who ever bore them meekly in the presence of all whom God had afflicted:—

> "In those days
> No knight of Arthur's noblest dealt in scorn;
> But, if a man were halt or hunch'd, in him
> By those whom God had made full-limb'd and tall,
> Scorn was allowed as part of his defect,
> And he was answered softly by the King
> And all his Table."

Of some poor ladies who were rescued by this doughty champion Boswell and others have left us quaint records. First in the list we have Mrs. Anna Williams, the daughter of a Welsh physician, who "having come to London in hopes of being cured of a cataract in both her eyes, which afterwards ended in total blindness, was kindly received as a constant visitor at his house while Mrs. Johnson lived; and after her death, having come under his roof in order to have an operation upon her eyes performed with more comfort to her than in lodgings, she had an apartment from him during the rest of her life, at all times when he had a house." Mrs. Anna Williams is described as being "a woman of more than ordinary talents and literature"; but, like most afflicted people, she was often peevish and unreasonable, nor, so far as we are able to judge from the specimens given of her conversation and writings, were her mental endowments at all sufficient to compensate for these defects. Moreover, she had certain peculiarities which must have been very trying to a man of Johnson's fastidious tastes. Let us take an instance of these from Boswell:—

"We went home to his house to tea. Mrs. Williams made it with sufficient dexterity, notwithstanding her blindness, though her manner of satisfying herself that the cups were full enough was a little awkward; she put her finger down a certain way, till she felt the tea touch it. In my first elation at being allowed the privilege of attending Dr. Johnson at his late visits to this lady, which was like being è secretioribus consiliis, I willingly drank cup after cup, as if it had been the Heliconian spring. But as the charm of novelty went off, I grew more fastidious; and besides I discovered that she was of a peevish temper."

Baretti also mentions that he did not relish dining at Johnson's, because he "hated to see the victuals paw'd by poor Mrs. Williams, that would often carve, though stone blind." How the Doctor would have stormed and scolded if this performance had taken place in other circumstances! "Madam," he would probably have said, "though partial to the flavour of both tea and meats, I am not a cannibal, and I object to the taste of human fingers," but before this poor afflicted lady he was dumb, swilling his tea, or gobbling his food, with perfect resignation. It is by such small acts of self-sacrifice that we often make our most acceptable offerings at the divine altar; and it is in such trifling magnanimities that the true gentleman declares himself. Which shows to the best advantage, think you, the flippant young advocate who records his disgust at poor Mrs. Williams's efforts to preside at tea, or the grim old Doctor obstinately shutting his eyes to her movements, and coming to share the repast night after night? To Boswell she was ever a crumpled roseleaf. We find him again complaining that :—

"Mrs. Williams was very peevish; and I wondered at Johnson's patience with her now, as I had often done on similar occasions. The truth is that his humane consideration of the forlorn and indigent state in which this lady was left by her father induced him to treat her with the utmost tenderness, and even to be desirous of procuring her amusement, so as sometimes to inconvenience many of his friends by carrying her to their houses, where, from her manner of eating, in consequence of her blindness, she could not but offend the delicacy of people of nice sensations."

And of this we have full corroboration in a statement by an eye-witness that, one afternoon, "Dr. Johnson carried Mrs. Williams to a fashionable reception, where he took the utmost care of her, guiding her about, and treating her as a father would his daughter."

An amusing instance of the order in which our Doctor was kept by this lady is narrated by Boswell, who, upon calling to take him to Dilly's the publisher's, where both were engaged to dine, found him "covered with dust and making no preparations for going abroad"—

"'How is this, Sir? (said I). Don't you recollect that you are going to dine at Mr. Dilly's?' JOHNSON : 'Sir, I did not think of going to Dilly's ; it went out of my head. I have ordered dinner at home with Mrs. Williams.' BOSWELL : 'But, my dear Sir, you know you were engaged to Mr. Dilly, and I told him so. He will expect you, and will be much disappointed if you don't come.' JOHNSON : 'You must talk to Mrs. Williams about this.'

"Here was a sad dilemma. I found that what I was so confident I had secured would yet be frustrated. He had accustomed himself to show Mrs. Williams such a degree of humane attention as frequently imposed some restraint upon him; and I knew that, if she should be obstinate, he would not stir. I hastened downstairs to the blind lady's room, and told her I was in great uneasiness, for Dr. Johnson had engaged to me to dine this day at Mr. Dilly's; but that he had told me he had forgotten his engagement, and had ordered dinner at home. 'Yes, Sir (said she, pretty peevishly), Dr. Johnson is to dine at home.' 'Madam (said I), his respect for you is such that I know he will not leave you unless you absolutely desire it. But as you have so much of his company, I hope you will be good enough to forego it for a day, as Mr. Dilly is a very worthy man, has frequently had agreeable parties at his house for Dr. Johnson, and will be vexed if Dr. Johnson neglect him to-day. And then, Madam, be pleased to consider my situation; I carried the message, and I assured Mr. Dilly that Dr. Johnson was to come, and no doubt he has made a dinner, and invited a company, and boasted of the honour he expected to have. I shall be quite disgraced if the Doctor is not there.' She gradually softened to my solicitations, which were certainly as earnest as most entreaties to ladies upon any occasion, and was graciously pleased to empower me to tell Dr. Johnson 'That, all things considered, she thought he should certainly go.' I flew back to him, still in dust, and careless of what should be the event, indifferent in his choice to go or stay; but as soon as I had announced to him Mrs. Williams's consent, he roared, 'Frank, a clean shirt,' and was very soon drest."

Upon similar occasions, when forced to dine away from home, it was the Doctor's custom to ask Mrs. Williams if she would like to have some dainty dish at her own dinner; and when she had made her choice, to order it to be sent in to her from a tavern, whilst on the way to his host's, so that she might as far as possible partake in his enjoyment. Truly, a beneficent Doctor, and none the less so because he was fully awake to her little unamiabilities. We are only made aware of this, however, by an occasional uncensuring remark, such as, "Mrs. Williams is in the country to try if she can improve her health ; she is very ill. Matters have come so about that she is in the country with very good accommodation ; but age and sickness and pride have made her so peevish that I was forced to bribe the maid to stay with her by a secret stipulation of half-a-crown a week over her wages." * Not a word of complaint, it may be observed. And so things went on, the poor lady growing more and more peevish every day, while he "supplied her, so far as could be desired, with all conveniences to make her excursion and abode pleasant and useful." He writes to Dr. Brocklesby—"Be so kind as to continue your

* Letter to Boswell, July 22, 1777.

attention to Mrs. Williams ; it is great consolation to the well, and still greater to the sick, that they find themselves not neglected " ; and when that gentleman sent him the news of her death he was greatly affected, declaring that his house had now become " a solitude." The old lady had contrived to save a little money out of a small pittance of her own, aided by a small yearly allowance from Mrs. Montagu ; and this she left, not to the man who had befriended her, but to a charity school.

There was another old lady in Johnson's queer establishment with whom we are told that Mrs. Williams used to "quarrel incessantly ; but as they can both be occasionally of service to each other, and as neither of them has any other place to go to, their animosity does not force them to separate." This was a Mrs. Desmoulins, widow of a writing-master, and daughter of a " Dr. Swinfen of Lichfield," who had been Johnson's godfather. Not only she, but her daughter, was taken in by the benevolent sage ; and " such was his humanity," says Boswell, " that Mrs. Desmoulins herself told me he allowed her half-a-guinea a week. Let it be remembered that this was above a twelfth part of his pension." Mrs. Desmoulins had the chief management of the Doctor's kitchen ; but her duties cannot have been very onerous, as he once confessed—" Our roasting is not magnificent, for we have no jack." MRS. THRALE : " No jack ! Why, how do they manage without ?" DR. JOHNSON : " Small joints, I believe, they manage with a string, and larger are done at the tavern. I have once thought (with a profound gravity) of buying a jack, because I think a jack is some credit to a house."

Besides the three ladies we have mentioned, there was yet another inmate of the establishment, a Miss Carmichael, also called " Poll " ; and a very disagreeable person she seems to have been, if we may judge from the following conversation :—

MR. THRALE: " But pray, Sir, who is the Poll you talk of? She that you used to abet in her quarrels with Mrs. Williams, and call out ' At her again, Poll ! Never flinch, Poll ' ? "

Dr. Johnson: "Why, I took to my Poll very well at first, but she won't do upon a nearer examination."

Mrs. Thrale: "How came she among you, Sir?"

Dr. Johnson: "Why I don't rightly remember; but we could spare her very well from us. Poll is a stupid slut; I had some hopes of her at first, but when I talked to her tightly and closely, I could make nothing of her; she was wiggle-waggle, and I could never persuade her to be categorical." *

It is by no means unlikely that many others, besides those of which we have the names, were accommodated under the Doctor's charitable roof. "Mrs. Thrale," says Miss Burney, "has often acquainted me that his house is quite filled and over-run with all sorts of strange creatures, whom he admits for mere charity, and because no one else will admit them—for his charity is unbounded—or rather, bounded only by his circumstances"; and this to the great disgust of the permanent inmates, Mesdames Williams, Desmoulins, etc., who, as

* Madame d'Arblay.

we have already mentioned, were by no means unready to vent their spleen upon the person responsible for these incursions. "Pray, Sir," asked Mrs. Thrale, "how does Mrs. Williams like all this tribe?" "Madam," replied the Doctor, "she does not like them at all; but their fondness for her is no greater." In fact, as he confessed, "Discord kept her residence at his habitation"; so that Mrs. Thrale tells how "he really was oftentimes afraid of going home because he was so sure to be met at the door with numberless complaints, and he used to lament pathetically to me that they made his life miserable from the impossibility he found in making them happy. If, however, I ventured to blame their ingratitude and censure their conduct, he would instantly set about softening the one and justifying the other, and finished commonly by telling me that I knew not how to make allowances for situations I had never experienced." When friends asked him how he could endure to be surrounded by such disagreeable people, he would reply—"If I did not assist them, no one else would, and they must be lost for want."* Sometimes there were lulls in the storm—times when, as he expressed it, "We have much malice, but no mischief"; but these halcyon periods, being due generally to the temporary absence of one or more of the combatants, were not entirely satisfactory to Johnson, who missed the din of voices. The stillness seemed to him unnatural. Thus we find him writing, "I have, by the migration of one of my ladies, more peace at home; but I remember an old savage chief that says of the Romans with great indignation: *Ubi solitudinem faciunt pacem appellant.*" Probably the worst time for the wretched man was when the "seraglio" were too indisposed to be capable of hostilities, yet would not dispense with their lord's companionship. "I found him," says Boswell, "one evening in Mrs. Williams's room, at tea and coffee with her and Mrs. Desmoulins, who were both ill; it was a sad scene, and he was not in very good humour." Poor old Doctor! The sick ladies were boring him horribly, and probably my Lady Lucan, or Mrs.

* Hawkins.

Vesey, or that giddy Miss Monckton, were at that moment awaiting him in their bright, cheerful drawing-rooms, where he knew he would be petted and flattered and feasted and have matters all his own way, yet he will not leave these two ailing old bodies to drink their tea and coffee in solitude ; he will sit with them and listen to their complaints and wheezings and, probably, their reproaches! It is not agreeable to him—far from it—and he feels slightly grumpy, but still he will bear them company as cheerfully as may be ; and when that young Boswell looks in, he can relieve his mind by abusing him, which he does.

Apart from what may be termed his domestic benevolence, we have many instances of Johnson's kindness, sought or unsought. Thus we find him writing to Mr. William Drummond as follows :—

"I must take the liberty of engaging you in an office of charity. Mrs. Heeley, the wife of Mr. Heeley, who had lately some office in your theatre, is my near relation, and now in great distress. They wrote me word of their situation some time ago, to which I returned them an answer which raised hopes of more than it is proper for me to give them . . . . I must beg, Sir, that you will enquire after them, and let me know what is to be done. I am willing to go ten pounds, and will transmit you such a sum, if upon examination you find it likely to be of use. If they are in immediate want, advance them what you think proper."

Ten pounds was no trifle to Johnson, who had to make a very small income go a very long way; but it was always his wont, were he ever so poorly off himself, to have something to spare for others. Whilst still a young and struggling man, with circumstances much embarrassed, we find him taking upon himself the payment of a debt due by his mother, and this without even acquainting her of the fact. "I look upon this," he writes to her creditor, "and on the future interest of that mortgage, as my own debt; and beg that you will give me directions how to pay it, and not mention it to my dear mother." *
But he was not, like some other generous people, at all given to reciprocity in this

* To Mr. Levett, in Lichfield, December 1, 1743.

respect. He shrank from receiving assistance even from his intimate friends. Thus, in a letter to one of the Miss Thrales, he writes: "A friend, whose name I will tell when your mamma has tried to guess it, sent to my physician to enquire whether this long train of illness had brought me into difficulties for want of money, with an invitation to send to him for what occasion required. I shall write this night to thank him, having no need to borrow." In connection with this may be mentioned one occasion upon which he must have been sorely tempted to depart from his usual practice. One night, in Fleet Street, "a gentlewoman (said he) begged I would give her my arm to assist her in crossing the street, which I accordingly did; upon which she offered me a shilling, supposing me to be the watchman. I perceived that she was somewhat in liquor." Whether he took the proffered guerdon, or not, is left to the imagination.

It was the nature of the man to stand by the weak and suffering in their affliction; to give them love and comfort when others would have avoided their unlively companionship. Pretty or ugly, young or old, high or low, in sickness or in health, woman was ever the object of his unswerving tenderness. Is there, in the whole range of fiction, a more touching description than the matter-of-fact one given below of his parting with an old acquaintance, then at the point of death?—

"Sunday, Oct. 18, 1797. Yesterday, Oct. 17, at about ten in the morning, I took my leave for ever of my dear old friend Catherine Chambers, who came to live with my mother about 1724, and has been but little parted from us since. . . . I desired all to withdraw, then told her that we were to part for ever: that, as Christians, we should part with prayer, and that I would, if she were willing, say a short prayer beside her. She expressed a great desire to hear me; and held up her poor hands, as she lay in bed, with great fervour, while I prayed, kneeling by her. . . . I then kissed her. She told me that to part was the greatest pain that she had ever felt, and that she hoped we would meet again in a better place. I expressed, with swelled eyes and great emotion of tenderness, the same hopes. We kissed and parted—I humbly hope to meet again, and to part no more."

Such was Johnson's knight-errantry—steadfast, catholic, pure. It may be wanting in picturesqueness; the knight is not young or debonnair, nor are the ladies, in sooth, very attractive; but the true spirit of chivalry is there notwithstanding, and a thoughtful artist might haply find a subject worthy of his brush in the list of Johnson's feats. May we venture to furnish him with a specimen?* The scene should be a night one, laid amid the grim streets of a sleeping city. A swinging oil-lamp sheds its dismal glow upon a strange figure, huge and misshapen, trudging sturdily over the rough pavement, one hand grasping a stout staff wherewith to support the rather lurching gait. Over the broad shoulders hangs a drooping figure, limp and bedraggled, clad in tawdry finery, on whose pallid face, half-shrouded by dishevelled hair, is the stamp of want, suffering, shame. The coffee-stall keeper, who is serving a wretched group of men and girls by such poor light as his paper-shade "glim" affords, looks up and touches his hat as the huge

* *Vide* pp. 23, 24.

Dr. Johnson as a Knight-Errant. 243

figure with its odd burden stalks by. A pair of roystering young Temple bucks on their way home to chambers pause to look after the old man, with the horse-laugh arrested on their very lips. The night-watchman seems about to challenge this apparition, but thinks better of it. Dawn is just breaking in the distance, and—that's all!

"THE LOST SHEEP."